FTCE Chemistry 6–12

Mil Adams

This page is intentionally left blank.

This page is intentionally left blank.

Table of Content

This page is intentionally left blank.

Chapter 1 – Questions

QUESTION 1

Which of the following best exemplifies the nature of science in chemistry?

  A. Scientific theories are absolute truths.
  B. Scientific knowledge in chemistry is static and unchanging.
  C. New evidence and data can lead to the revision of scientific theories.
  D. Scientists rely solely on intuition and personal beliefs to make discoveries.

Answer:

QUESTION 2

A chemistry teacher is discussing a controversial scientific topic related to a chemical process with the class. Some students express their viewpoints based on personal beliefs rather than scientific evidence. How can the teacher encourage critical thinking and scientific reasoning in such discussions?

  A. Discourage students from sharing personal beliefs and focus solely on the scientific evidence presented.
  B. Acknowledge and respect diverse viewpoints, but emphasize the importance of evidence-based arguments.
  C. Censor any discussion that involves personal beliefs to maintain a strictly scientific environment.
  D. Ignore personal beliefs and assert the teacher's viewpoint as the only valid perspective.

Answer:

QUESTION 3

Which of the following best describes the process of osmosis?

  A. The movement of water molecules from a region of lower solute concentration to a region of higher solute concentration through a semipermeable membrane.
  B. The movement of solute molecules from a region of higher concentration to a region of lower concentration through a semipermeable membrane.
  C. The movement of water molecules from a region of higher solute concentration to a region of lower solute concentration through a semipermeable membrane.
  D. The movement of solvent molecules from a region of higher pressure to a region of lower pressure through a semipermeable membrane.

Answer:

QUESTION 4

What is the term for a substance that speeds up a chemical reaction without being consumed in the process?

  A. Catalyst
  B. Inhibitor
  C. Reactant
  D. Product

Answer:

QUESTION 5

Why is it essential for chemistry teachers to promote scientific literacy among their students?

  A. To encourage students to pursue careers in chemistry.
  B. To equip students with the ability to critically evaluate scientific information.
  C. To memorize chemical formulas and equations more effectively.
  D. To decrease the time spent on practical laboratory experiments.

Answer:

QUESTION 6

A chemistry teacher wants to determine the concentration of an unknown solution. Which technique is suitable for this purpose?

    A.   Titration.
    B.   Distillation.
    C.   Filtration.
    D.   Decantation.

Answer:

QUESTION 7

A student is studying the effects of concentration on reaction rate. The student conducts an experiment using a fixed amount of reactants and varies the concentration of one reactant. The student observes that the reaction rate increases with increasing concentration of the reactant. Based on these results, which of the following conclusions can be drawn?

    A.   Increasing the concentration of the reactant slows down the reaction rate.
    B.   Decreasing the concentration of the reactant speeds up the reaction rate.
    C.   The reaction rate is not affected by the concentration of the reactant.
    D.   Increasing the concentration of the reactant increases the reaction rate

Answer:

QUESTION 8

Which of the following chemical reactions is an example of an endothermic reaction?

    A.   Combustion of methane (CH_4) in the presence of oxygen to produce carbon dioxide (CO_2) and water (H_2O).
    B.   The dissolution of ammonium nitrate (NH_4NO_3) in water.
    C.   The reaction between sodium hydroxide (NaOH) and hydrochloric acid (HCl) to form sodium chloride (NaCl) and water (H_2O).
    D.   The decomposition of calcium carbonate ($CaCO_3$) into calcium oxide (CaO) and carbon dioxide (CO_2).

Answer:

QUESTION 9

Which of the following molecules is an example of a polar covalent compound?

    A.   Oxygen gas (O_2).
    B.   Methane (CH_4).
    C.   Carbon dioxide (CO_2).
    D.   Water (H_2O).

Answer:

QUESTION 10

A chemistry teacher is explaining the concept of chemical equilibrium to students. Which of the following examples best represents a dynamic equilibrium in a chemical reaction?

    A.   The burning of a candle.
    B.   The dissolution of sugar in water.
    C.   The evaporation of water from a puddle.
    D.   The forward and reverse reaction rates of a reversible reaction becoming equal.

Answer:

QUESTION 11

A chemistry teacher wants to compare the effectiveness of three different catalysts in a chemical reaction. What study design is suitable for this investigation?

    A.   Case study.
    B.   Double-blind study.
    C.   Controlled experiment.
    D.   Longitudinal study.

Answer:

QUESTION 12

Which of the following is NOT a property of acids?

    A.   Sour taste.
    B.   Turns blue litmus paper red.
    C.   Reacts with metals to produce hydrogen gas.
    D.   Increases the pH of a solution.

Answer:

QUESTION 13

A chemistry student is investigating the effect of temperature on the solubility of a solid substance in water. The student measures the mass of the substance that dissolves in 100 mL of water at different temperatures and plots the data on a graph. The graph shows that the solubility of the substance increases as the temperature rises. Which of the following statements provides a possible explanation for this trend?

    A.   The molar mass of the substance increases with temperature.
    B.   The water molecules gain kinetic energy at higher temperatures.
    C.   The intermolecular forces between the substance and water weaken at higher temperatures.
    D.   The density of water decreases with temperature, allowing more solid to dissolve.

Answer:

QUESTION 14

A student conducted an experiment to investigate the effect of temperature on the reaction rate. They observed that as the temperature increased, the reaction rate also increased. Based on this observation, what conclusion can be drawn?

    A.   Increasing temperature causes the reaction to slow down.
    B.   Increasing temperature has no effect on the reaction rate.
    C.   The student made an error in recording the data.
    D.   Increasing temperature accelerates the reaction rate.

Answer:

QUESTION 15

Which of the following substances is considered an allotrope of carbon?

    A.   Methane (CH_4).
    B.   Ethanol (C_2H_5OH).
    C.   Diamond.
    D.   Sodium chloride ($NaCl$).

Answer:

QUESTION 16

A student performs an experiment to determine the pH of a solution using litmus paper. The litmus paper turns blue, indicating that the solution is basic. What additional information does this observation provide?

    A. The exact pH value of the solution.
    B. The concentration of the basic solution.
    C. The identity of the solute in the solution.
    D. The molarity of the basic solution.

Answer:

QUESTION 17

In a lab experiment, two chemistry students were tasked with determining the identity of an unknown white powder. Student A used a spectroscope to analyze the light emitted when the powder was heated, while Student B performed a series of chemical reactions with the powder to observe its behavior. Both students correctly identified the unknown powder as sodium chloride (table salt). Which student's method is more reliable, and why?

    A. Student A's method is more reliable because spectroscopy provides precise and definitive results.
    B. Student B's method is more reliable because chemical reactions offer more conclusive evidence.
    C. Both methods are equally reliable because they both led to the correct identification.
    D. Neither method is reliable, and the experiment should be repeated using a different approach.

Answer:

QUESTION 18

A chemistry teacher assigns a group of students to investigate the properties of a new chemical compound. The students conduct their experiments and observe unexpected results that do not align with their initial hypothesis. What should the students do next to uphold the principles of scientific inquiry?

    A. Discard the results and repeat the experiments until the expected outcomes are obtained.
    B. Modify the data to match the initial hypothesis to avoid inconsistencies.
    C. Analyze the unexpected results, consider alternative explanations, and revise their hypothesis accordingly.
    D. Ignore the discrepancies and proceed with publishing their findings as planned.

Answer:

QUESTION 19

A chemistry teacher wants to demonstrate the reaction between an acid and a metal to her students. She has a sample of hydrochloric acid (HCl) and two metal samples, zinc (Zn) and copper (Cu). The teacher places both metal samples separately in the same volume of hydrochloric acid and observes the reactions. To her surprise, only one metal reacts with the acid. Which metal reacts, and what might be the reason behind this observation?

    A. Zinc (Zn) reacts with the acid because it is a transition metal.
    B. Copper (Cu) reacts with the acid because it has a higher atomic number.
    C. Zinc (Zn) reacts with the acid because it is more electronegative.
    D. Copper (Cu) reacts with the acid because it forms a stronger metallic bond.

Answer:

QUESTION 20

In the context of chemistry education, how does technology enhance the learning experience?

  A. By replacing traditional laboratory experiments with virtual simulations.
  B. By reducing the need for critical thinking in problem-solving.
  C. By increasing the focus on theoretical concepts rather than practical applications.
  D. By providing ready-made answers to complex chemical equations.

Answer:

QUESTION 21

Which method is commonly used to determine the concentration of an acid in a solution?

  A. Chromatography.
  B. Spectroscopy.
  C. Titration.
  D. Electrolysis.

Answer:

QUESTION 22

When designing an experiment to investigate a chemical reaction, why is it essential to include control groups?

  A. Control groups provide additional data points to increase the size of the dataset.
  B. Control groups help in making predictions about future experiments.
  C. Control groups serve as a comparison to assess the specific effects of the manipulated variables.
  D. Control groups are necessary to prove the researcher's initial hypothesis.

Answer:

QUESTION 23

A chemistry class is conducting an experiment to investigate the effect of a catalyst on the rate of a chemical reaction. They perform the reaction in three separate trials: one without a catalyst, one with a solid catalyst, and one with a dissolved catalyst. Surprisingly, they find that the reaction rate remains the same in all three trials. How can this unexpected result be explained?

  A. The catalyst was not properly prepared or added to the reaction mixture.
  B. The concentration of the catalyst was too high, leading to an equilibrium state.
  C. The reaction is not catalyzed by the tested catalysts.
  D. The reaction rate may be limited by factors other than the presence of a catalyst.

Answer:

QUESTION 24

An atom with 7 protons, 7 neutrons, and 7 electrons undergoes beta decay. Which of the following statements is correct about the resulting atom?

  A. The number of protons increases, the number of neutrons decreases, and the number of electrons remains the same.
  B. The number of protons decreases, the number of neutrons increases, and the number of electrons remains the same.
  C. The number of protons remains the same, the number of neutrons decreases, and the number of electrons increases.
  D. The number of protons remains the same, the number of neutrons increases, and the number of electrons decreases.

Answer:

QUESTION 25

A gas occupies a volume of 4 liters at 27°C and 1 atmosphere pressure. If the temperature is raised to 127°C, what will be the new volume of the gas assuming constant pressure?

A. 1 liter.
B. 4 liters.
C. 8 liters.
D. 16 liters.

Answer:

QUESTION 26

The element with the highest ionization energy among the following is:

A. Sodium (Na)
B. Magnesium (Mg)
C. Aluminum (Al)
D. Silicon (Si)

Answer:

QUESTION 27

When comparing two substances with the same mass, the one with a higher specific heat capacity will:

A. Have a higher temperature change when heated.
B. Have a lower temperature change when heated.
C. Release more energy when cooled.
D. Absorb more energy when cooled.

Answer:

QUESTION 28

What type of energy change occurs when a solid solute dissolves in a solvent?

A. Endothermic
B. Exothermic
C. Isobaric
D. Isometric

Answer:

QUESTION 29

A chemical reaction has a positive enthalpy change, indicating an endothermic process. However, it occurs spontaneously and rapidly at room temperature. How is this possible?

A. The reaction has a high activation energy.
B. The reaction has a low activation energy.
C. The reaction is catalyzed.
D. The reaction is at equilibrium.

Answer:

QUESTION 30

Which of the following elements is likely to have similar chemical properties and reactivity?

    A.   Sodium (Na)
    B.   Iron (Fe)
    C.   Argon (Ar)
    D.   Phosphorus (P)

Answer:

QUESTION 31

A gas sample is compressed at constant temperature. How does the average kinetic energy of the gas molecules change during this process?

    A.   It increases.
    B.   It decreases.
    C.   It remains constant.
    D.   It fluctuates randomly.

Answer:

QUESTION 32

Which statement about electron energy levels is true?

    A.   Electrons in higher energy levels are closer to the nucleus than electrons in lower energy levels.
    B.   Electrons in higher energy levels have lower energy than electrons in lower energy levels.
    C.   Electrons in higher energy levels have higher energy than electrons in lower energy levels.
    D.   Electrons in different energy levels have the same energy.

Answer:

QUESTION 33

Who proposed the first modern atomic theory with the concept of indivisible atoms?

    A.   John Dalton
    B.   Niels Bohr
    C.   Dmitri Mendeleev
    D.   Ernest Rutherford

Answer:

QUESTION 34

In a spontaneous process, which quantity always increases?

    A.   Gibbs free energy (ΔG)
    B.   Enthalpy (ΔH)
    C.   Entropy (ΔS)
    D.   Internal energy (ΔU)

Answer:

QUESTION 35

The enthalpy diagram for a chemical reaction shows a significant increase in potential energy for the products compared to the reactants. What can be said about this reaction?

    A. It is endothermic and requires energy input.
    B. It is exothermic and releases energy.
    C. It is a phase transition.
    D. It involves no energy changes.

Answer:

QUESTION 36

What is the molecular geometry of a molecule with the electron domain geometry of trigonalbipyramidal and two lone pairs?

    A. Linear
    B. Bent
    C. T-shaped
    D. Square pyramidal

Answer:

QUESTION 37

The modern periodic table is arranged based on:

    A. Atomic mass
    B. Number of protons
    C. Electron configuration
    D. Atomic number

Answer:

QUESTION 38

Which of the following interactions is responsible for holding the protons and neutrons together in the atomic nucleus?

    A. Electromagnetic force
    B. Gravitational force
    C. Strong nuclear force
    D. Weak nuclear force

Answer:

QUESTION 39

What is the primary factor that determines the phase of a substance at a given temperature and pressure?

    A. The density of the substance.
    B. The mass of the substance.
    C. The arrangement of molecules in the substance.
    D. The color of the substance.

Answer:

QUESTION 40

What is the maximum number of electrons that can occupy the 4d sublevel?

A. 2
B. 8
C. 10
D. 18

Answer:

QUESTION 41

According to the kinetic molecular theory, gas pressure is caused by:

A. The weight of gas molecules
B. The constant collisions of gas particles with each other
C. The attractive forces between gas molecules
D. The volume occupied by gas molecules

Answer:

QUESTION 42

Which type of calorimetry measures the heat exchanged at constant volume?

A. Isoperiboliccalorimetry
B. Adiabatic calorimetry
C. Bomb calorimetry
D. Isothermal calorimetry

Answer:

QUESTION 43

Which type of reaction releases heat to the surroundings and decreases the enthalpy of the system?

A. Endothermic reaction
B. Exothermic reaction
C. Isobaric reaction
D. Isometric reaction

Answer:

QUESTION 44

What is the correct IUPAC name for the compound with the molecular formula CH_3COOH?

A. Acetic acid
B. Ethanol
C. Propanoic acid
D. Formic acid

Answer:

QUESTION 45

Which group of elements on the periodic table is known for having a complete outer electron shell (octet) and are generally unreactive (chemically inert)?

A. Alkali metals
B. Noble gases
C. Halogens
D. Transition metals

Answer:

QUESTION 46

Which of the following gases will have the highest average velocity at the same temperature?

A. Nitrogen (N2)
B. Oxygen (O2)
C. Carbon dioxide (CO2)
D. Hydrogen (H2)

Answer:

QUESTION 47

Which of the following statements is true about isotopes of the same element?

A. They have different atomic numbers.
B. They have different chemical properties.
C. They have different numbers of protons.
D. They have different numbers of electrons.

Answer:

QUESTION 48

The ideal gas law, PV = nRT, relates the pressure, volume, temperature, and amount of gas with the gas constant (R). Which units are used for the gas constant (R)?

A. L·atm/(mol·K)
B. g/L
C. mol/L
D. J/(mol·K)

Answer:

QUESTION 49

A student is conducting an experiment to measure the heat capacity of an unknown metal. The student heats the metal sample to a high temperature and then transfers it to a calorimeter containing water at room temperature. After a while, the student records the final temperature of the water and the metal. However, the student forgot to take into account the heat gained or lost by the calorimeter itself. How would this oversight affect the calculated heat capacity of the unknown metal?

A. The calculated heat capacity would be too high.
B. The calculated heat capacity would be too low.
C. The calculated heat capacity would be accurate.
D. The calculated heat capacity would depend on the metal's specific heat capacity.

Answer:

QUESTION 50

Which of the following processes is characterized by an increase in entropy?

    A. Freezing of water
    B. Condensation of steam
    C. Evaporation of a liquid
    D. Formation of a crystal lattice

Answer:

QUESTION 51

What is the molecular geometry of a molecule with the electron domain geometry of tetrahedral and no lone pairs?

    A. Trigonal pyramidal
    B. Linear
    C. Tetrahedral
    D. Trigonal planar

Answer:

QUESTION 52

Which of the following elements is a metalloid?

    A. Silicon (Si)
    B. Potassium (K)
    C. Chlorine (Cl)
    D. Iron (Fe)

Answer:

QUESTION 53

The kinetic molecular theory explains which of the following properties of gases?

    A. The specific heat capacity of gases.
    B. The compressibility of gases.
    C. The electrical conductivity of gases.
    D. The density of gases.

Answer:

QUESTION 54

An atom has 15 protons, 15 neutrons, and 16 electrons. What is the net charge of this atom?

    A. +1
    B. -1
    C. 0
    D. +2

Answer:

QUESTION 55

Which law of thermodynamics defines the concept of entropy and the natural direction of processes towards increased disorder?

- A. First Law of Thermodynamics
- B. Second Law of Thermodynamics
- C. Third Law of Thermodynamics
- D. Zeroth Law of Thermodynamics

Answer:

QUESTION 56

A chemistry student performs an experiment to investigate the dissolution of a solid in water. They measure the change in enthalpy (ΔH) of the process to be positive and the change in entropy (ΔS) to be negative. Based on these results, is the process spontaneous at all temperatures, and what can be said about the dissolving solid's solubility?

- A. The process is spontaneous at all temperatures, and the solid is highly soluble.
- B. The process is non-spontaneous at all temperatures, and the solid is insoluble.
- C. The process is spontaneous only at high temperatures, and the solid is moderately soluble.
- D. The process is non-spontaneous at all temperatures, and the solid's solubility cannot be determined.

Answer:

QUESTION 57

What type of energy change occurs when a gas undergoes an adiabatic expansion?

- A. Endothermic
- B. Exothermic
- C. Isobaric
- D. Adiabatic

Answer:

QUESTION 58

Two isomers have the same molecular formula and the same arrangement of atoms but differ in the spatial arrangement of groups around a chiral center. What type of isomers are they?

- A. Structural isomers
- B. Geometric isomers
- C. Conformational isomers
- D. Optical isomers

Answer:

QUESTION 59

Which functional group is present in an aldehyde?

- A. Aldehyde
- B. Carboxylic acid
- C. Ketone
- D. Amine

Answer:

QUESTION 60

Two molecules, A and B, have similar molecular weights and identical molecular formulas but different structures. Molecule A has a linear structure, while Molecule B has a bent structure. Which type of intermolecular forces is likely to be stronger between Molecules A and B?

A. Dipole-dipole interactions
B. London dispersion forces
C. Hydrogen bonding
D. Ion-dipole interactions

Answer:

QUESTION 61

Which type of chemical bond involves the attraction between a partially positive hydrogen atom and a highly electronegative atom in another molecule?

A. Covalent bond
B. Ionic bond
C. Metallic bond
D. Hydrogen bond

Answer:

QUESTION 62

Which of the following statements is true about the equilibrium constant (Kc) for an exothermic reaction?

A. Kc increases with increasing temperature.
B. Kc decreases with increasing temperature.
C. Kc is not affected by changes in temperature.
D. Kc depends on the concentration of reactants only.

Answer:

QUESTION 63

Which of the following is an example of a buffer solution?

A. Pure water
B. A solution of strong acid and strong base
C. A solution of weak acid and its conjugate base
D. A solution of a highly concentrated acid

Answer:

QUESTION 64

Which of the following conditions will increase the voltage output of a galvanic cell?

A. Decreasing the concentration of the electrolyte.
B. Decreasing the surface area of the electrodes.
C. Increasing the distance between the electrodes.
D. Using electrodes with higher standard reduction potentials.

Answer:

QUESTION 65

Which type of chemical bond involves the sharing of electrons between two non-metal atoms?

    A. Covalent bond
    B. Ionic bond
    C. Metallic bond
    D. Hydrogen bond

Answer:

QUESTION 66

A student performs an experiment and observes that a solid sample of Compound X does not conduct electricity, both in its pure form and when dissolved in water. However, the student notices that when Compound X is dissolved in an ionic liquid, it becomes a good conductor of electricity. Which type of compound is most likely Compound X?

    A. Ionic compound
    B. Nonpolar covalent compound
    C. Polar covalent compound
    D. Metallic compound

Answer:

QUESTION 67

When two atoms form a covalent bond, the bond's strength is influenced by various factors. Which factor does NOT directly affect the bond strength?

    A. The number of shared electrons
    B. The distance between the bonded atoms
    C. The size of the atoms involved
    D. The electronegativity difference between the atoms

Answer:

QUESTION 68

Which of the following statements is true about the equilibrium constant (Kc) for a reversible chemical reaction?

    A. Kc is affected by the initial concentrations
    B. Kc is the same for the forward and reverse reactions.
    C. Kc changes if the temperature is increased.
    D. Kc is a measure of the rate of the reaction.

Answer:

QUESTION 69

What is the relationship between the pH and pOH of a solution at 25°C?

    A. pH + pOH = 14
    B. pH × pOH = 1
    C. pH - pOH = 0
    D. pH ÷ pOH = 2

Answer:

Which of the following statements is true regarding the Nernst equation?

- A. It relates the standard reduction potentials of two half-reactions.
- B. It calculates the equilibrium constant (K) of a redox reaction.
- C. It describes the relationship between the cell potential and the reaction quotient (Q).
- D. It is applicable only to galvanic cells and not to electrolytic cells.

Answer:

QUESTION 71

Which statement accurately represents an endothermic chemical reaction?

- A. The reaction releases energy in the form of heat.
- B. The products have higher energy than the reactants.
- C. The reaction causes an increase in the surrounding temperature.
- D. The reaction absorbs energy from the surroundings.

Answer:

QUESTION 72

Among the intermolecular forces, which one is the weakest but is responsible for the state of matter known as "gas" at room temperature for many substances?

- A. Dipole-dipole interactions
- B. London dispersion forces
- C. Hydrogen bonding
- D. Ion-dipole interactions

Answer:

QUESTION 73

A compound is found to be a good conductor of electricity in both its solid state and when dissolved in water. Its bonding involves sharing electrons between two elements with significantly different electronegativities. Which type of bonding is most likely present in this compound?

- A. Covalent bonding
- B. Ionic bonding
- C. Metallic bonding
- D. Van der Waals forces

Answer:

QUESTION 74

The concept of bond enthalpy is used to measure the strength of a chemical bond. Which statement best describes the relationship between bond enthalpy and bond strength?

- A. Higher bond enthalpy corresponds to stronger bonds.
- B. Bond enthalpy is inversely proportional to bond strength.
- C. Bond enthalpy is directly proportional to the number of covalent bonds in a molecule.
- D. Stronger bonds have lower bond enthalpy values.

Answer:

QUESTION 75

The equilibrium constant (Kc) for the reaction: $2SO_2(g) + O_2(g) \rightleftharpoons 2SO_3(g)$ is 500. What can be said about the position of the equilibrium?

A. The equilibrium favors the reactants.
B. The equilibrium favors the products.
C. The equilibrium is at a balance with nearly equal concentrations of reactants and products.
D. It is impossible to determine the equilibrium position from the given information.

Answer:

QUESTION 76

Which of the following indicates that a titration is complete during an acid-base titration experiment?

A. A sudden change in color of the indicator solution.
B. The pH of the solution becomes neutral (pH 7).
C. The volume of the titrant added is equal to the volume of the analyte solution.
D. The appearance of bubbles in the solution.

Answer:

QUESTION 77

What is the primary purpose of a fuel cell?

A. To produce electrical energy from the combustion of fuels.
B. To store energy in the form of chemical bonds.
C. To convert mechanical energy into electrical energy.
D. To generate electrical energy through electrochemical reactions.

Answer:

QUESTION 78

Which of the following factors affects the rate of a chemical reaction?

A. The chemical formula of the reactants.
B. The total mass of the reactants.
C. The volume of the reaction vessel.
D. The concentration of the reactants.

Answer:

QUESTION 79

Which factor affects the strength of hydrogen bonding between molecules?

A. Molecular weight of the molecules
B. Dipole moment of the molecules
C. Number of atoms in the molecules
D. Presence of lone pairs of electrons in the molecules

Answer:

QUESTION 80

When atoms form a chemical bond, which factor(s) influence the strength of the bond?

    A.   The distance between the nuclei of the bonded atoms
    B.   The atomic number of the elements involved
    C.   The number of valence electrons in the elements
    D.   The boiling point of the elements

Answer:

QUESTION 81

Which of the following factors would NOT influence the position of a chemical equilibrium?

    A.   Temperature
    B.   Pressure
    C.   Catalyst concentration
    D.   Initial reactant concentrations

Answer:

QUESTION 82

The equilibrium constant (Kc) for the reaction: $2NO(g) + Cl_2(g) \rightleftharpoons 2NOCl(g)$ is 0.05. If the concentration of NOCl is increased while the concentrations of NO and Cl2 remain unchanged, what happens to the equilibrium position?

    A.   Shift towards NO and Cl2
    B.   Shift towards NOCl
    C.   No change
    D.   Insufficient data to determine

Answer:

QUESTION 83

A chemistry teacher has a solution with a pH of 9. What is the pOH value of this solution?

    A.   5
    B.   7
    C.   9
    D.   14

Answer:

QUESTION 84

Which of the following processes is responsible for the spontaneous generation of an electric current in a galvanic cell?

    A.   The migration of ions through a salt bridge.
    B.   The flow of electrons in the external circuit.
    C.   The dissociation of reactants in the electrolyte.
    D.   The diffusion of gases between the half-cells.

Answer:

QUESTION 85

Which type of bond is the strongest?

    A. Covalent bond
    B. Ionic bond
    C. Metallic bond
    D. Hydrogen bond

Answer:

QUESTION 86

Which of the following statements is true regarding energy changes in chemical bonding?

    A. Breaking a covalent bond releases energy, while forming a covalent bond requires energy.
    B. Breaking an ionic bond requires energy, while forming an ionic bond releases energy.
    C. Breaking a metallic bond requires energy, while forming a metallic bond releases energy.
    D. Breaking a hydrogen bond releases energy, while forming a hydrogen bond requires energy.

Answer:

QUESTION 87

The equilibrium constant (Kc) for the following reaction:

$2NO_2(g) \rightleftharpoons N_2O_4(g)$

is 0.25. If the concentration of N2O4 is doubled while the concentration of NO2 remains the same, what happens to the equilibrium position?

    A. Shift towards N2O4
    B. Shift towards NO2
    C. No change
    D. Insufficient data to determine

Answer:

QUESTION 88

For the reaction: $N_2O_4(g) \rightleftharpoons 2NO_2(g)$, an increase in temperature favors the endothermic direction. How can you restore the original position of the equilibrium if heat is added to the system?

    A. Increase the volume of the container.
    B. Decrease the volume of the container.
    C. Add more N2O4 to the system.
    D. Add more NO2 to the system.

Answer:

QUESTION 89

A chemistry teacher is titrating a strong acid with a strong base. Initially, the pH of the acid solution is 1. As the titrant (base) is added, the pH increases gradually and then suddenly rises sharply to pH 7. What does this sharp rise in pH indicate during the titration?

A. The titration is incomplete, and more titrant needs to be added.
B. The equivalence point has been reached.
C. The pH meter is malfunctioning and giving inaccurate readings.
D. The pH of the solution has become neutral.

Answer:

QUESTION 90

Consider the following electrochemical cell:

Pb(s) | Pb^2+(aq) || Cu^2+(aq) | Cu(s)

If the concentration of Cu^2+ ions in the Cu^2+(aq) half-cell is increased while keeping all other conditions constant, how will this affect the cell potential?

A. The cell potential will increase.
B. The cell potential will decrease.
C. The cell potential will remain unchanged.
D. The cell potential will depend on the temperature.

Answer:

QUESTION 91

A student sets up an electrochemical cell with a copper electrode immersed in a solution containing copper ions and a silver electrode immersed in a solution containing silver ions. The student then connects the electrodes externally with a wire. After some time, the copper electrode gains mass, while the silver electrode loses mass. Explain the observations and the underlying electrochemical processes.

A. The copper electrode acts as the anode and undergoes oxidation, losing electrons, while the silver electrode acts as the cathode and undergoes reduction, gaining electrons.
B. The copper electrode acts as the cathode and undergoes reduction, gaining electrons, while the silver electrode acts as the anode and undergoes oxidation, losing electrons.
C. Both the copper and silver electrodes act as anodes, undergoing oxidation and losing electrons.
D. Both the copper and silver electrodes act as cathodes, undergoing reduction and gaining electrons.

Answer:

QUESTION 92

Which statement best describes the concept of bond polarity in a molecule?

A. Bond polarity is determined by the number of atoms involved in forming a bond.
B. Bond polarity refers to the strength of the bond between two atoms in a molecule.
C. Bond polarity results from an unequal sharing of electrons between atoms, leading to regions of partial positive and partial negative charges.
D. Bond polarity only occurs in ionic compounds.

Answer:

QUESTION 93

What is the effect of adding a catalyst to a chemical reaction at equilibrium?

    A.   It shifts the equilibrium position towards the products.
    B.   It shifts the equilibrium position towards the reactants.
    C.   It increases the value of the equilibrium constant (Kc).
    D.   It has no effect on the equilibrium position or the value of Kc.

Answer:

QUESTION 94

Which of the following statements accurately defines a base according to its behavior and how it is defined?

    A.   A base is a substance that donates protons (H^+ ions) in a chemical reaction.
    B.   A base is a substance that accepts protons (H^+ ions) in a chemical reaction.
    C.   A base is a substance that increases the concentration of hydroxide ions (OH^-) in a solution.
    D.   A base is a substance that decreases the concentration of hydronium ions (H_3O^+) in a solution.

Answer:

QUESTION 95

Consider the following oxidation-reduction reaction:

$2K(s) + Cl2(g) -> 2KCl(s)$

Identify the reducing agent in this reaction.

    A.   K(s)
    B.   Cl2(g)
    C.   KCl(s)
    D.   There is no reducing agent in this reaction.

Answer:

QUESTION 96

A student conducts an electrolysis experiment using a solution of sodium chloride (NaCl) and observes that chlorine gas (Cl2) is produced at one electrode. What is the likely identity of the electrode where chlorine gas is being evolved?

    A.   Copper electrode (Cu)
    B.   Silver electrode (Ag)
    C.   Platinum electrode (Pt)
    D.   Aluminum electrode (Al)

Answer:

QUESTION 97

What is the role of energy in chemical reactions?

    A.   Energy is neither released nor absorbed during reactions.
    B.   Energy is created during all reactions.
    C.   Energy is absorbed to break bonds and released when new bonds form.
    D.   Energy is released to break bonds and absorbed when new bonds form.

Answer:

QUESTION 98

In a chemical reaction, the reactants have an activation energy that must be overcome to form the products. What role does the catalyst play in this process?

A. A catalyst lowers the activation energy required for the reaction to proceed.
B. A catalyst increases the activation energy required for the reaction to proceed.
C. A catalyst provides additional energy to the reactants, increasing the activation energy.
D. A catalyst has no effect on the activation energy of the reaction.

Answer:

QUESTION 99

A chemistry teacher is investigating the effect of a catalyst on the rate of a chemical reaction. The teacher observes that adding the catalyst significantly increases the reaction rate. Which of the following statements best describes the role of the catalyst in this reaction?

A. The catalyst provides additional energy to the reactants.
B. The catalyst lowers the activation energy of the reaction.
C. The catalyst increases the concentration of reactants.
D. The catalyst introduces new reaction intermediates.

Answer:

QUESTION 100

Which of the following factors would NOT affect the rate of dissolution of a solid solute in a liquid solvent?

A. Particle size of the solute.
B. Temperature of the solvent.
C. Pressure applied to the system.
D. Polarity of the solute.

Answer:

QUESTION 101

For the reaction between ethane (C_2H_6) and oxygen (O_2) to produce carbon dioxide (CO_2) and water (H_2O), which of the following statements is TRUE?

A. Ethane is the limiting reactant if an equal number of moles of C_2H_6 and O_2 are used.
B. Oxygen is the limiting reactant if an equal number of moles of C_2H_6 and O_2 are used.
C. The limiting reactant can be determined without knowing the balanced equation.
D. The reaction cannot occur as written; it needs a different stoichiometry.

Answer:

QUESTION 102

Which of the following statements is true about metallic bonding?

A. It involves the transfer of electrons from one atom to another.
B. It forms compounds with distinct chemical formulas.
C. It allows for high electrical conductivity.
D. It only occurs between nonmetals.

Answer:

QUESTION 103

In a reaction, 2 moles of A react with 3 moles of B to produce 1 mole of C and 1 mole of D. What is the limiting reactant, and how many moles of the excess reactant remain after the reaction?

    A.   A is the limiting reactant, and 1 mole of B remains.
    B.   A is the limiting reactant, and 0.5 moles of B remains.
    C.   B is the limiting reactant, and 1 mole of A remains.
    D.   B is the limiting reactant, and 0.5 moles of A remains.

Answer:

QUESTION 104

Mr. Davis is planning a scientific investigation with his chemistry class. He wants to ensure that students learn to identify potential sources of error in their experiments. Which approach would be most effective for achieving this objective?

    A.   Providing students with a detailed procedure to follow step-by-step during the investigation.
    B.   Encouraging students to use the same tools and materials for all experiments to ensure consistency.
    C.   Having students collaborate and discuss their findings with peers during the investigation.
    D.   Guiding students in identifying and discussing possible sources of error after completing the investigation.

Answer:

QUESTION 105

When a student mixes two clear colorless solutions in a test tube and observes a solid precipitate forming, which type of chemical reaction is most likely occurring?

    A.   Combustion reaction
    B.   Single displacement reaction
    C.   Double displacement reaction
    D.   Decomposition reaction

Answer:

QUESTION 106

A chemistry teacher is studying the reaction kinetics of a complex chemical reaction. After analyzing the rate data, the teacher observes that the reaction rate increases with increasing temperature. However, at extremely high temperatures, the reaction rate starts to decrease. What phenomenon might be causing this unexpected behavior?

    A.   Deactivation of the catalyst at high temperatures.
    B.   The reactants become more stable at higher temperatures.
    C.   The reaction has reached equilibrium at extremely high temperatures.
    D.   The activation energy of the reaction changes with temperature.

Answer:

QUESTION 107

Which of the following statements is true about the Tyndall effect?

    A.   It is observed only in solutions.
    B.   It is observed in both solutions and suspensions.
    C.   It is observed only in suspensions.
    D.   It is observed in colloids and suspensions but not in solutions.

Answer:

QUESTION 108

What approximate volume of 0.5 M hydrochloric acid (HCl) is required to react completely with 25 grams of calcium hydroxide (Ca(OH)2)? (Molar mass: Ca(OH)2 = 74.09 g/mol)

A.   35 mL
B.   70 mL
C.   140 mL
D.   280 mL

Answer:

QUESTION 109

Which type of bond is characterized by the transfer of electrons from one atom to another?

A.   Covalent bond
B.   Hydrogen bond
C.   Ionic bond
D.   Metallic bond

Answer:

QUESTION 110

During a laboratory experiment, a student obtained a percent yield of 80% for a chemical reaction. If the theoretical yield was 50 grams, what was the actual yield obtained by the student?

A.   20 grams
B.   40 grams
C.   50 grams
D.   64 grams

Answer:

QUESTION 111

Mr. Johnson is teaching a unit on the properties of matter. He wants his students to make systematic observations and measurements to understand these properties fully. Which of the following activities would be most effective in achieving his objective?

A.   Asking students to read a chapter in the textbook about the properties of matter.
B.   Providing students with a variety of substances and asking them to identify their states of matter and measure their mass and volume.
C.   Showing students a video about the properties of matter and discussing it afterward.
D.   Giving students a quiz on the definitions of different properties of matter.

Answer:

QUESTION 112

A chemistry teacher is conducting an experiment to investigate the factors that influence the rate of a chemical reaction. She keeps the concentration of reactants constant but varies the temperature. Which concept of collision theory is she testing?

A.   Activation energy
B.   Orientation factor
C.   Effective collisions
D.   Temperature dependence

Answer:

QUESTION 113

A chemistry teacher is studying the reaction between two substances, A and B, and finds that the reaction rate increases with increasing concentration of both A and B. However, when a third substance, C, is added to the reaction mixture, the rate remains constant regardless of the concentrations of A, B, and C. What could be the role of substance C in this reaction?

A. Substance C acts as a catalyst, speeding up the reaction without being consumed.
B. Substance C reacts with A and B to form a stable product, slowing down the reaction.
C. Substance C binds to A and B, preventing them from reacting and slowing down the rate.
D. Substance C alters the temperature of the reaction mixture, compensating for changes in reactant concentration.

Answer:

QUESTION 114

Which of the following chemical equations is balanced correctly?

A. 2H2 + O2 → 2H2O
B. CH4 + 2O2 → CO2 + 2H2O
C. NaCl + H2SO4 → Na2SO4 + 2HCl
D. N2 + 3H2 → 2NH3

Answer:

QUESTION 115

A compound with the empirical formula CH_2O has a molar mass of 180 g/mol. What is its molecular formula?

A. $C_2H_4O_2$
B. $C_4H_8O_4$
C. $C_6H_{12}O_6$
D. $C_3H_6O_3$

Answer:

QUESTION 116

How does the electronegativity difference between two bonded atoms influence the bond type?

A. It has no effect on the bond type.
B. Higher electronegativity difference leads to a weaker bond.
C. Higher electronegativity difference leads to a stronger ionic bond.
D. Higher electronegativity difference leads to a stronger covalent bond.

Answer:

QUESTION 117

A chemical reaction is exothermic and releases 100 kJ of energy. If the reaction is performed in a calorimeter, what will be the effect on the temperature of the surroundings?

A. The temperature of the surroundings will increase.
B. The temperature of the surroundings will decrease.
C. The temperature of the surroundings will remain constant.
D. The effect on the temperature of the surroundings cannot be determined without additional information.

Answer:

QUESTION 118

Mr. Lee is introducing the concept of chemical equations to his chemistry class. He wants to guide his students in making connections with their prior knowledge to understand the new topic better. Which of the following approaches would best help Mr. Lee achieve this goal?

    A. Presenting a list of chemical equations for students to memorize.
    B. Showing a video demonstration of chemical reactions and asking students to write down their observations.
    C. Engaging students in a class discussion about common household chemical reactions they might have encountered before.
    D. Assigning a worksheet with a set of chemical equations to balance.

Answer:

QUESTION 119

A chemistry teacher is investigating the reaction rate of a chemical reaction in the presence of a catalyst. After analyzing the data, the teacher notices that the reaction rate remains unchanged, regardless of the concentration of the catalyst used. What could be a plausible explanation for this observation?

    A. The catalyst is not functioning properly.
    B. The reaction is a zero-order reaction.
    C. The catalyst is inhibiting the reaction.
    D. The concentration of reactants is too low.

Answer:

QUESTION 120

Which of the following statements correctly defines a solution?

    A. A heterogeneous mixture of two or more substances.
    B. A mixture in which one substance is uniformly dispersed in another substance at the molecular level.
    C. A mixture with large particles that settle out over time.
    D. A mixture of two or more elements in fixed proportions.

Answer:

QUESTION 121

In a chemical reaction, 8 grams of magnesium (Mg) reacts with 16 grams of oxygen (O2) to produce magnesium oxide (MgO). What is the limiting reactant in this reaction?

    A. Magnesium (Mg)
    B. Oxygen (O2)
    C. Both reactants are in the same ratio; none is limiting.
    D. Not enough information to determine.

Answer:

QUESTION 122

A compound contains 20% hydrogen, 25% carbon, and 55% oxygen by mass. The molar mass of the compound is 120 g/mol. What is the molecular formula of the compound?

    A. $C_4H_6O_4$
    B. $C_6H_{12}O_6$
    C. $C_3H_6O_3$
    D. $C_9H_{18}O_9$

Answer:

QUESTION 123

When two atoms form a covalent bond, what determines the bond's strength and stability?

    A. The number of electrons involved in the bond.
    B. The size of the atoms involved.
    C. The overlap of atomic orbitals.
    D. The overall size of the molecule.

Answer:

QUESTION 124

What is the purpose of using a buffer solution in a chemical laboratory or industrial process?

    A. To increase the reaction rate of a chemical reaction.
    B. To reduce the concentration of reactants in a solution.
    C. To maintain a stable pH of a solution when small amounts of acid or base are added.
    D. To increase the solubility of a solute in a solvent.

Answer:

QUESTION 125

Mr. Rodriguez is planning a lesson on chemical bonding for his chemistry class. He wants to sequence the learning activities to challenge his students and expand their understanding of the topic. Which of the following activities would be the most appropriate final step in his lesson plan?

    A. Presenting a lecture on the different types of chemical bonds and their properties.
    B. Giving students a set of chemical formulas and asking them to identify the type of chemical bond in each compound.
    C. Conducting a lab activity where students can observe the formation of ionic and covalent compounds.
    D. Assigning a worksheet with problems related to Lewis structures and bonding.

Answer:

QUESTION 126

During a chemical reaction, the concentration of reactants decreases as the reaction proceeds. Which of the following rate expressions correctly represents this situation for a general reaction "A + B → Products"?

    A. Rate = k[A][B]
    B. Rate = k[A]^2[B]^2
    C. Rate = k[B]
    D. Rate = k[A]

Answer:

QUESTION 127

Which factor would most likely increase the solubility of a gas in a liquid?

    A. Decreasing the temperature.
    B. Increasing the pressure.
    C. Decreasing the pressure.
    D. Increasing the volume.

Answer:

QUESTION 128

The combustion of methane (CH4) produces carbon dioxide (CO2) and water (H2O) according to the balanced equation:

CH4 + 2O2 → CO2 + 2H2O

What is the percent yield if 8 grams of methane are combusted, producing 12 grams of carbon dioxide, and 4 grams of water?

A. 50%
B. 75%
C. 80%
D. 100%

Answer:

QUESTION 129

Which of the following compounds has the highest percent composition of nitrogen by mass?

A. NH_4NO_3 (Ammonium nitrate)
B. NH_3 (Ammonia)
C. NO_2 (Nitrogen dioxide)
D. N_2O (Nitrous oxide)

Answer:

QUESTION 130

The bond length between two atoms in a covalent bond is influenced by:

A. The number of lone pairs around the central atom.
B. The number of atoms in the molecule.
C. The strength of the bond.
D. The size of the bonded atoms.

Answer:

QUESTION 131

What is the molality of a solution containing 30 grams of potassium chloride (KCl) dissolved in 500 grams of water?

A. 0.8 mol/kg
B. 1. 2 mol/kg
C. 3 mol/kg
D. 4 mol/kg

Answer:

QUESTION 132

Mr. Roberts is planning a lesson on the periodic table for his chemistry class. He wants to sequence the learning activities to uncover common misconceptions and build upon his students' prior knowledge. Which of the following activities would be the most appropriate first step in his lesson plan?

A. Providing students with a blank periodic table and asking them to fill in the elements' symbols and atomic masses.
B. Giving a lecture on the history of the periodic table and its organization.
C. Conducting a lab activity where students can observe the properties of different elements.
D. Asking students to read a chapter in the textbook about the periodic table.

Answer:

QUESTION 133

A chemistry teacher is comparing two different reactions, P and Q. Reaction P has a lower activation energy than reaction Q. How will this difference in activation energies impact the reactions?

    A.   Reaction P will be slower than reaction Q.
    B.   Both reactions will proceed at the same rate.
    C.   Reaction P will be faster than reaction Q.
    D.   Reaction P will be reversible, while reaction Q will be irreversible.

Answer:

QUESTION 134

Colligative properties of solutions depend on:

    A.   The nature of the solute particles only.
    B.   The nature of the solvent particles only.
    C.   The total number of solute particles.
    D.   The concentration of the solvent.

Answer:

QUESTION 135

Which of the following reactions represents an exothermic process?

    A. $N_2(g) + 3H_2(g) \rightarrow 2NH_3(g)$
    B. $CaCO_3(s) \rightarrow CaO(s) + CO_2(g)$
    C. $2H_2O(l) \rightarrow 2H_2(g) + O_2(g)$
    D. $CH_4(g) + 2O_2(g) \rightarrow CO_2(g) + 2H_2O(g)$

Answer:

QUESTION 136

A sample of unknown compound X has a mass of 0.45 grams. When the sample is completely combusted, it produces 1.56 grams of carbon dioxide (CO_2) and 0.45 grams of water (H_2O). What is the empirical formula of compound X?

    A. CH
    B. CH_2O
    C. $C_2H_2O_2$
    D. $C_3H_3O_3$

Answer:

QUESTION 137

Which of the following statements best explains why stoichiometry is essential for understanding chemical reactions?

    A.   Stoichiometry allows us to determine the energy changes during a reaction.
    B.   Stoichiometry helps to predict the equilibrium position of a reaction.
    C.   Stoichiometry provides a quantitative relationship between reactants and products in a chemical reaction.
    D.   Stoichiometry is essential for understanding the kinetics of a reaction.

Answer:

QUESTION 138

Mr. Thompson is preparing a chemistry lesson for his diverse classroom, considering students' interests, knowledge, understanding, abilities, and experiences. What would be the most inclusive approach to selecting instructional materials for this lesson?

A. Using materials that cater to a single learning style.
B. Incorporating a variety of materials to address different learning preferences.
C. Focusing on materials that challenge the students' existing knowledge.
D. Relying solely on a single textbook for consistency.

Answer:

QUESTION 139

Ms. Martinez is planning a lesson on the states of matter for her chemistry class. She wants to sequence the learning activities to challenge her students and expand their understanding of the topic. Which of the following activities would be the most appropriate final step in her lesson plan?

A. Giving students a worksheet with problems related to the behavior of particles in different states of matter.
B. Providing a set of true/false statements about the properties of different states of matter for students to discuss.
C. Conducting a lab activity where students can observe changes in states of matter under different conditions.
D. Presenting a lecture on the intermolecular forces involved in different states of matter.

Answer:

QUESTION 140

A chemistry teacher is comparing two different chemical reactions, R and S. Reaction R is a first-order reaction, while reaction S is a second-order reaction. How will their reaction rates change with respect to changes in reactant concentration?

A. Doubling the concentration of reactants will double the rate of both reactions.
B. Doubling the concentration of reactants will quadruple the rate of reaction R and double the rate of reaction S.
C. Doubling the concentration of reactants will double the rate of reaction R and quadruple the rate of reaction S.
D. Doubling the concentration of reactants will have no effect on the rate of either reaction.

Answer:

QUESTION 141

Which concentration unit is expressed as moles of solute per liter of solution?

A. Molarity (M).
B. Molality (m).
C. Mass percent.
D. Parts per million (ppm).

Answer:

QUESTION 142

What type of bond is formed between two atoms when they share a pair of electrons?

A. Ionic bond
B. Covalent bond
C. Metallic bond
D. Hydrogen bond

Answer:

QUESTION 143

Mr. Smith has just completed teaching a unit on chemical reactions. He wants to gauge his students' understanding of the concepts learned throughout the unit. Which assessment method would be most suitable for this purpose?

    A.   Asking students to write a reflective essay on their learning journey during the unit.
    B.   Conducting a formal written exam with multiple-choice and short-answer s.
    C.   Using peer assessment, where students evaluate each other's understanding of chemical reactions.
    D.   Having students maintain field journals of their observations during lab experiments.

Answer:

QUESTION 144

This question is intentionally removed.

QUESTION 145

Which of the following chemical equations represents a net ionic equation for a precipitation reaction?

    A.   AgNO3 + NaCl → AgCl + NaNO3
    B.   HCl + NaOH → NaCl + H2O
    C.   FeCl3 + 3NaOH → Fe(OH)3 + 3NaCl
    D.   KOH + HNO3 → KNO3 + H2O

Answer:

QUESTION 146

A compound is analyzed and found to contain 30% carbon, 6.67% hydrogen, and 63.33% oxygen by mass. The molar mass of the compound is approximately 90 g/mol. What is the molecular formula of the compound?

    A. $C_3H_6O_3$
    B. $C_6H_{12}O_6$
    C. $C_6H_6O_3$
    D. $C_6H_{12}O_3$

Answer:

QUESTION 147

The reaction between potassium hydroxide (KOH) and sulfuric acid (H2SO4) can be represented by the following balanced equation:

2KOH + H2SO4 → K2SO4 + 2H2O

What approximate mass of potassium sulfate (K2SO4) is formed when 20 grams of potassium hydroxide (KOH) reacts with excess sulfuric acid (H2SO4)? (Molar mass: KOH = 56.11 g/mol, H2SO4 = 98.08 g/mol, K2SO4 = 174.26 g/mol)

    A.   15.0 g
    B.   31.0 g
    C.   58.0 g
    D.   116.0 g

Answer:

QUESTION 148

Which of the following best describes the scientific attitude that chemistry teachers should instill in their students?

   A.   Accepting scientific claims without ing.
   B.   Relying on personal experiences rather than data.
   C.   Encouraging skepticism and critical thinking.
   D.   Ignoring experimental errors to support a hypothesis.

Answer:

QUESTION 149

A group of students is investigating the reaction between a metal and an acid. They tested three different metals (A, B, and C) with the same concentration of acid and observed the following:

Metal A reacted vigorously, producing a large amount of gas bubbles.

Metal B reacted mildly, producing some gas bubbles.

Metal C showed no visible reaction with the acid.

Which of the following explanations could best account for these observations?

   A.   Metal A is the most reactive metal because it produced the most gas bubbles.
   B.   Metal C is the least reactive metal because it showed no visible reaction.
   C.   Metal B is the least reactive metal because it produced only a moderate amount of gas bubbles.
   D.   The reaction rate of each metal with the acid cannot be determined based on the given observations.

Answer:

QUESTION 150

What is the term for the amount of heat required to raise the temperature of one gram of a substance by one degree Celsius?

   A.   Enthalpy
   B.   Entropy
   C.   Specific heat capacity
   D.   Heat of fusion

Answer:

QUESTION 151

Which of the following scenarios best demonstrates the application of chemistry in everyday life?

   A.   A chemist synthesizing a new drug in a research laboratory.
   B.   A chemist analyzing soil samples to assess environmental pollution.
   C.   A chef preparing a meal using fresh ingredients.
   D.   A student conducting experiments in a high school chemistry lab.

Answer:

QUESTION 152

When recording measurements in the laboratory, what is the correct way to express uncertainty?

A. Round down to the nearest whole number.
B. Round up to the nearest whole number.
C. Include all decimal places shown on the instrument.
D. Use the instrument's uncertainty value.

Answer:

QUESTION 153

Which of the following materials is characterized as radioactive?

A. Carbon-12.
B. Oxygen-16.
C. Uranium-238.
D. Nitrogen-14.

Answer:

QUESTION 154

The teacher asks the students to identify the type of reaction for each equation. Which of the following statements best explains the classification of the reactions?

A. Reaction 1 is a decomposition reaction because one reactant breaks down into two products.
B. Reaction 1 is a combination reaction because two reactants combine to form one product.
C. Reaction 2 is a double displacement reaction because two compounds exchange ions to form new compounds.
D. Reaction 2 is a displacement reaction because a metal displaces another metal in a compound.

Answer:

QUESTION 155

Which type of chemical reaction involves the exchange of ions between two reacting compounds, resulting in the formation of two new compounds?

A. Double displacement reaction
B. Decomposition reaction
C. Combustion reaction
D. Single displacement reaction

Answer:

QUESTION 156

A chemistry teacher demonstrates an exothermic reaction to the class. After the reaction is completed, the teacher touches the reaction vessel and finds it warm to the touch. One student asks, "If the reaction released energy, why don't we feel a sudden burst of heat during the reaction?" Which of the following responses best addresses the student's ?

A. "The released heat is absorbed by the surroundings before we can feel it."
B. "The heat energy released during the reaction is converted into light energy, not heat."
C. "The heat is converted into kinetic energy, making the particles move faster."
D. "The heat released during the reaction is too small to be felt by our skin."

Answer:

QUESTION 157

A chemistry teacher is discussing the concept of pH with students. One student asks why a solution with a pH of 3 is considered more acidic than a solution with a pH of 5. What would be the most appropriate response by the teacher?

A. "The solution with a pH of 3 has more hydrogen ions (H+) than the solution with a pH of 5."
B. "The solution with a pH of 3 has fewer hydroxide ions (OH-) than the solution with a pH of 5."
C. "The solution with a pH of 3 has a higher concentration of water (H2O) than the solution with a pH of 5."
D. "The solution with a pH of 3 has a higher concentration of carbon dioxide (CO2) than the solution with a pH of 5."

Answer:

QUESTION 158

When analyzing scientific data, what is the purpose of creating a data table?

A. To showcase the researchers' experimental skills.
B. To summarize the main findings of the study.
C. To present data in an organized and structured manner.
D. To provide additional information not included in the report.

Answer:

QUESTION 159

What is the primary reason for the use of radioisotopes in various applications, such as medical imaging and cancer treatment?

A. They are abundant and readily available.
B. They have longer half-lives than stable isotopes.
C. They emit harmful radiation that destroys cancer cells.
D. They undergo radioactive decay, emitting detectable radiation.

Answer:

QUESTION 160

What is the role of statistical analysis in chemistry research?

A. To manipulate data to support desired outcomes.
B. To provide a way to hide experimental errors.
C. To draw valid conclusions and quantify uncertainties.
D. To replace the need for conducting experiments.

Answer:

QUESTION 161

Which of the following is an example of a redox reaction (oxidation-reduction reaction)?

A. The burning of wood in the presence of oxygen.
B. The dissolution of sugar in water.
C. The melting of ice.
D. The mixing of oil and water.

Answer:

QUESTION 162

What is the fundamental principle behind the modern periodic table?

- A. Atomic mass
- B. Atomic number
- C. Number of electrons
- D. Number of protons

Answer:

QUESTION 163

A chemistry teacher conducts an experiment to investigate the rate of reaction between two substances, X and Y. After collecting the data, the teacher notices that doubling the concentration of substance X results in a four-fold increase in the rate of reaction. Which statement best describes the reaction order with respect to substance X?

- A. The reaction order with respect to substance X is zero.
- B. The reaction order with respect to substance X is one.
- C. The reaction order with respect to substance X is two.
- D. The reaction order with respect to substance X is four.

Answer:

QUESTION 164

A chemistry teacher is conducting a demonstration to explain the concept of solubility to students. The teacher adds excess solid sugar to a container of water and stirs it until no more sugar dissolves. Which of the following statements best explains why the remaining sugar does not dissolve?

- A. The sugar particles have decreased in size and cannot dissolve further.
- B. The water has reached its maximum volume and cannot dissolve more sugar.
- C. The water temperature has decreased, lowering the solubility of sugar.
- D. The solution has reached the maximum amount of sugar that can dissolve at that temperature.

Answer:

QUESTION 165

What is the purpose of a control variable in a scientific experiment?

- A. To provide an additional experimental group for comparison.
- B. To maintain constant conditions throughout the experiment.
- C. To ensure that the experiment is conducted with minimal resources.
- D. To randomize the assignment of participants to different groups.

Answer:

QUESTION 166

In what ways can chemistry teachers encourage creativity and innovation among their students?

- A. By providing step-by-step instructions for all experiments.
- B. By discouraging students from ing established scientific principles.
- C. By assigning repetitive and mundane laboratory tasks.
- D. By allowing freedom in experimental design and exploration of new ideas.

Answer:

QUESTION 167

A chemistry class is studying the properties of acids and bases. The teacher performs an experiment where a few drops of a universal indicator are added to three different solutions: a strong acid, a strong base, and a neutral solution. The results show that the strong acid solution turns bright red, the strong base solution turns deep blue, and the neutral solution turns green. Which statement best explains these color changes?

A. The universal indicator changes color based on the concentration of hydrogen ions (H+) in the solution.
B. The universal indicator reacts with the acid and base to form different colored compounds.
C. The universal indicator is sensitive to the temperature of the solutions, leading to varying colors.
D. The universal indicator is not suitable for differentiating between acids and bases.

Answer:

QUESTION 168

What is the first step in designing a scientific investigation?

A. Collecting data from previous studies.
B. Formulating a hypothesis.
C. Analyzing the results.
D. Selecting the appropriate measurement tools.

Answer:

QUESTION 169

A chemistry class conducted an experiment to determine the solubility of a solid in water at different temperatures. The students noticed that as they increased the temperature, the solubility of the solid also increased. One student suggested that this relationship would be the same for all solids dissolved in water. Is the student's suggestion valid?

A. Yes, solubility is always directly proportional to temperature for all solids in water.
B. No, solubility can increase or decrease depending on the specific solid and the solvent.
C. Yes, the student's suggestion is correct for ionic compounds only.
D. No, solubility remains constant regardless of temperature for all solids.

Answer:

QUESTION 170

Why is it crucial for chemistry teachers to address the potential bias in scientific research?

A. Bias is an inherent part of scientific investigations that cannot be eliminated.
B. Acknowledging bias undermines the credibility of the research findings.
C. Bias can influence the design, execution, and interpretation of experiments.
D. Addressing bias in research papers increases the length of the publication.

Answer:

QUESTION 171

Which of the following statements is true regarding exergonic and endergonic reactions?

A. Exergonic reactions require energy input to proceed, while endergonic reactions release energy.
B. Exergonic reactions release energy, while endergonic reactions require energy input to proceed.
C. Exergonic reactions and endergonic reactions are the same and can be used interchangeably.
D. Exergonic reactions and endergonic reactions are not involved in any biological processes.

Answer:

QUESTION 172

How does the history of chemistry contribute to the current understanding of the discipline?

    A.   It provides entertaining stories about past discoveries.
    B.   It highlights the mistakes made by early chemists.
    C.   It offers valuable lessons about the limitations of early instruments and techniques.
    D.   It shows that chemistry has not changed significantly over time.

Answer:

QUESTION 173

Why is it essential to use controls in a scientific investigation?

    A.   Controls help ensure safety during the experiment.
    B.   Controls add complexity to the experiment.
    C.   Controls provide a basis for comparison to the experimental group.
    D.   Controls are required by law in most experiments.

Answer:

QUESTION 174

A chemistry teacher wants to determine the concentration of an unknown acid solution. The teacher performs a titration with a standardized base solution and records the volume of base solution needed to neutralize the acid. Which additional information is necessary to calculate the concentration of the unknown acid?

    A.   The temperature during the titration.
    B.   The color change of the indicator used in the titration.
    C.   The concentration of the standardized base solution.
    D.   The chemical formula of the unknown acid.

Answer:

QUESTION 175

Which of the following is a characteristic of a homogenous mixture?

    A.   It consists of visibly distinct phases.
    B.   It has variable composition throughout.
    C.   It has uniform properties throughout.
    D.   It scatters light, making the mixture opaque

Answer:

QUESTION 176

Which historical model of atomic structure proposed that atoms consist of tiny indivisible particles and do not contain any subatomic particles?

    A.   Plum Pudding Model
    B.   Dalton's Atomic Theory
    C.   Rutherford Model
    D.   Bohr Model

Answer:

QUESTION 177

A hypothetical element, Element X, is located in Group 17 of the periodic table. Its electron configuration is [Kr] $5s^2$ $4d^{10}$ $5p^5$. What chemical property is characteristic of Element X, and why is it significant?

    A.   Element X is a noble gas and is chemically inert due to a complete outer electron shell.
    B.   Element X is an alkali metal and is highly reactive due to a single valence electron.
    C.   Element X is a transition metal and can exhibit multiple oxidation states due to partially filled d orbitals.
    D.   Element X is a halogen and is likely to form ions with a charge of -1 due to a strong electron affinity.

Answer:

QUESTION 178

Which of the following statements accurately describes an isotope?

    A.   Isotopes are atoms of the same element with the same number of protons but different numbers of neutrons.
    B.   Isotopes are atoms of different elements with the same number of protons and neutrons.
    C.   Isotopes are atoms of the same element with different numbers of protons but the same number of neutrons.
    D.   Isotopes are atoms of different elements with different numbers of protons and neutrons.

Answer:

QUESTION 179

 Which of the following is an example of a chemical property of matter?

    A.   Melting point
    B.   Color
    C.   Flammability
    D.   Density

Answer:

QUESTION 180

If a chemical reaction is at equilibrium and the temperature is increased, what will be the effect on the position of the equilibrium?

    A.   Shift to the right, favoring the forward reaction.
    B.   Shift to the left, favoring the reverse reaction.
    C.   No effect on the equilibrium position.
    D.   The reaction will no longer be at equilibrium.

Answer:

QUESTION 181

During an exothermic reaction, what happens to the enthalpy of the products compared to the reactants?

    A.   Enthalpy decreases
    B.   Enthalpy increases
    C.   Enthalpy remains constant
    D.   Entropy is inversely proportional to enthalpy

Answer:

QUESTION 182

An atom of an element has 12 protons, 12 neutrons, and 11 electrons. What is the electrical charge of this atom?

    A. +1
    B. 0
    C. +11
    D. -1

Answer:

QUESTION 183

An ideal gas is initially at a temperature T. If the temperature of the gas is doubled while keeping the pressure constant, how does the volume change?

    A. It is halved.
    B. It is doubled.
    C. It is quadrupled.
    D. It remains constant.

Answer:

QUESTION 184

Which of the following elements has the highest electronegativity?

    A. Hydrogen (H)
    B. Carbon (C)
    C. Oxygen (O)
    D. Fluorine (F)

Answer:

QUESTION 185

The elements in the Periodic Table are arranged based on:

    A. Atomic number
    B. Atomic mass
    C. Number of protons
    D. Number of neutrons

Answer:

QUESTION 186

A chemical reaction has a negative value of ΔG. What can be said about the spontaneity of the reaction under standard conditions?

    A. The reaction is spontaneous at any temperature.
    B. The reaction is spontaneous only at low temperatures.
    C. The reaction is non-spontaneous at all temperatures.
    D. The reaction spontaneity depends on the activation energy.

Answer:

QUESTION 187

Which of the following elements has the highest electronegativity value?

    A.   Hydrogen (H)
    B.   Oxygen (O)
    C.   Fluorine (F)
    D.   Chlorine (Cl)

Answer:

QUESTION 188

Which statement is true about the relationship between pressure and volume of a gas at constant temperature?

    A.   Pressure is inversely proportional to volume.
    B.   Pressure is directly proportional to volume.
    C.   Pressure and volume have no relationship at constant temperature.
    D.   Pressure and volume follow a random pattern.

Answer:

QUESTION 189

The atomic number of an element is determined by the number of:

    A.   Protons in the nucleus.
    B.   Neutrons in the nucleus.
    C.   Electrons in the outermost energy level.
    D.   Electrons in the innermost energy level.

Answer:

QUESTION 190

Which law of thermodynamics states that energy cannot be created or destroyed, only converted from one form to another?

    A.   First Law of Thermodynamics
    B.   Second Law of Thermodynamics
    C.   Third Law of Thermodynamics
    D.   Zeroth Law of Thermodynamics

Answer:

QUESTION 191

Two chemical reactions, A and B, have the same negative value for ΔG under standard conditions. Reaction A occurs spontaneously at room temperature, while reaction B does not. What could be the reason for this difference in spontaneity?

    A.   Reaction A has a lower activation energy than reaction B.
    B.   Reaction A has a higher activation energy than reaction B.
    C.   Reaction A has a higher ΔH (enthalpy change) than reaction B.
    D.   Reaction A has a lower ΔH (enthalpy change) than reaction B.

Answer:

QUESTION 192

During a chemical reaction, the volume of the reaction mixture decreases, and the temperature of the surroundings increases. What can you infer about the nature of the reaction regarding its enthalpy change (ΔH) and the sign of work (w) done by the system on the surroundings?

A. The reaction is exothermic, and work is done on the system by the surroundings.
B. The reaction is endothermic, and work is done by the system on the surroundings.
C. The reaction is exothermic, and work is done by the system on the surroundings.
D. The reaction is endothermic, and work is done on the system by the surroundings.

Answer:

QUESTION 193

Which of the following statements about isotopes is correct?

A. Isotopes have the same number of protons and electrons but different numbers of neutrons.
B. Isotopes have the same number of protons and neutrons but different numbers of electrons.
C. Isotopes have the same number of neutrons and electrons but different numbers of protons.
D. Isotopes have the same number of protons but different numbers of neutrons and electrons.

Answer:

QUESTION 194

Which of the following is the correct electron configuration for a neutral nitrogen atom (atomic number 7)?

A. 1s2 2s2 2p4
B. 1s2 2s2 2p5
C. 1s2 2s1 2p4
D. 1s2 2s2 2p3

Answer:

QUESTION 195

Which form of energy is associated with the motion of atoms and molecules in a substance?

A. Potential energy
B. Kinetic energy
C. Thermal energy
D. Chemical energy

Answer:

QUESTION 196

A chemistry class performs an experiment to measure the enthalpy change (ΔH) of a chemical reaction using a coffee-cup calorimeter. The initial and final temperatures of the reaction mixture are recorded. However, during the experiment, some heat is lost to the surroundings due to poor insulation. How would this heat loss affect the calculated value of ΔH for the reaction?

A. The calculated ΔH would be lower than the actual value.
B. The calculated ΔH would be higher than the actual value.
C. The heat loss would not affect the calculated ΔH significantly.
D. The effect of heat loss on ΔH cannot be determined without additional information.

Answer:

QUESTION 197

A student investigates the dissolution of two different solid solutes in a solvent: Substance A and Substance B. Substance A dissolves endothermically, while Substance B dissolves exothermically. What factors might explain this difference in energy change?

A. The molecular weight of Substance A is greater than Substance B.
B. Substance A forms weaker intermolecular bonds with the solvent.
C. Substance B has a higher melting point than Substance A.
D. Substance B has a lower solubility in the solvent.

Answer:

QUESTION 198

Two isomers have the same molecular formula and the same arrangement of atoms but differ in rotation around a single bond. What type of isomers are they?

A. Structural isomers
B. Geometric isomers
C. Conformational isomers
D. Optical isomers

Answer:

QUESTION 199

What type of bond is primarily responsible for the unique properties of water, such as high surface tension and specific heat capacity?

A. Covalent bond
B. Ionic bond
C. Metallic bond
D. Hydrogen bond

Answer:

QUESTION 200

Two elements, Y and Z, have the same number of valence electrons and similar atomic sizes. Element Y forms a covalent bond with Element Z, but the resulting compound is a poor conductor of electricity. Which property of the elements could explain this behavior?

A. Electronegativity
B. Ionization energy
C. Electron affinity
D. Metallic character

Answer:

QUESTION 201

In some molecules, resonance structures can be drawn to represent the delocalization of electrons. Which of the following molecules exhibits resonance, and what does it imply about the actual structure of the molecule?

A. Carbon dioxide; The carbon-oxygen bonds are of equal length and strength.
B. Nitrate ion (NO_3^-); The nitrogen-oxygen bonds are single bonds.
C. Benzene (C_6H_6); The carbon-carbon bonds alternate between single and double bonds.
D. Water; The oxygen-hydrogen bonds are slightly polar.

Answer:

QUESTION 202

For the reaction: $2H_2O(g) \rightleftharpoons 2H_2(g) + O_2(g)$, if the pressure is increased, what will be the effect on the equilibrium position?

    A.   Shift towards H2O
    B.   Shift towards H2 and O2
    C.   Shift towards H2 and decrease towards O2
    D.   Shift towards O2 and decrease towards H2

Answer:

QUESTION 203

A chemistry teacher wants to prepare a buffer solution with a pH of 5. Which of the following combinations would be the most appropriate for preparing the buffer?

    A.   A weak acid with a pKa of 4 and its conjugate base with a pKb of 6.
    B.   A strong acid with a pKa of 2 and its conjugate base with a pKb of 9.
    C.   A weak acid with a pKa of 6 and its conjugate base with a pKb of 8.
    D.   A strong acid with a pKa of 1 and its conjugate base with a pKb of 13.

Answer:

QUESTION 204

Which functional group is present in an organic acid?

    A.   Aldehyde
    B.   Carboxylic acid
    C.   Ketone
    D.   Amine

Answer:

QUESTION 205

Which type of bond is formed between a metal and a non-metal and involves the pooling of valence electrons?

    A.   Covalent bond
    B.   Ionic bond
    C.   Metallic bond
    D.   Hydrogen bond

Answer:

QUESTION 206

Which type of chemical bond is formed when atoms share electrons, resulting in a localized electron density between the bonded atoms?

    A.   Ionic bond
    B.   Hydrogen bond
    C.   Covalent bond
    D.   Metallic bond

Answer:

QUESTION 207

The compound ammonia (NH3) and methane (CH4) have similar molecular shapes, yet ammonia exhibits hydrogen bonding, while methane does not. Why does ammonia show hydrogen bonding, while methane does not?

A. Ammonia has stronger covalent bonds than methane, allowing for hydrogen bonding.
B. Ammonia has a higher boiling point than methane, indicating the presence of hydrogen bonding.
C. The hydrogen atoms in ammonia have a higher electron affinity than in methane, promoting hydrogen bonding.
D. Hydrogen bonding is only observed in compounds containing nitrogen and hydrogen.

Answer:

QUESTION 208

Consider the reaction: $N_2(g) + 3H_2(g) \rightleftharpoons 2NH_3(g)$. If the volume of the reaction container is suddenly decreased by half while keeping the temperature constant, what will be the effect on the equilibrium position?

A. The equilibrium will shift towards N2 and H2.
B. The equilibrium will shift towards NH3.
C. The equilibrium will remain unchanged.
D. The equilibrium will shift towards the direction with more moles of gas.

Answer:

QUESTION 209

Which of the following is a characteristic of a strong base?

A. It partially dissociates in water, releasing only a few OH⁻ ions.
B. It completely dissociates in water, releasing many OH⁻ ions.
C. It has a high pH value.
D. It is a poor conductor of electricity in solution.

Answer:

QUESTION 210

In a redox reaction, which species is being oxidized when its oxidation number decreases?

A. The species with the lowest electronegativity.
B. The species with the highest electronegativity.
C. The reducing agent.
D. The species with the higher concentration.

Answer:

QUESTION 211

Which type of solid exhibits a regular arrangement of positive ions in a sea of delocalized electrons?

A. Ionic solid
B. Molecular solid
C. Covalent solid
D. Metallic solid

Answer:

QUESTION 212

Which of the following substances is most likely to exhibit hydrogen bonding?

A. CH4 (methane)
B. H2S (hydrogen sulfide)
C. NH3 (ammonia)
D. PH3 (phosphine)

Answer:

QUESTION 213

Which type of bond is present between two water molecules, creating the unique properties of water, such as high surface tension and cohesion?

A. Covalent bond
B. Ionic bond
C. Hydrogen bond
D. Metallic bond

Answer:

QUESTION 214

A chemical reaction has an equilibrium constant (Kc) value of 100. What does this value indicate about the reaction at equilibrium?

A. The reaction heavily favors the products.
B. The reaction heavily favors the reactants.
C. The reaction is almost at equilibrium with a balance between products and reactants.
D. The reaction does not reach equilibrium.

Answer:

QUESTION 215

Consider the reaction: CO(g) + 2H2(g) ⇌ CH3OH(g). If the pressure of the system is increased, what will be the effect on the equilibrium position?

A. The equilibrium will shift towards CO and H2.
B. The equilibrium will shift towards CH3OH.
C. The equilibrium will remain unchanged.
D. The equilibrium will shift towards the direction with more moles of gas.

Answer:

QUESTION 216

Which of the following statements accurately defines oxidation and reduction?

A. Oxidation is the gain of electrons, while reduction is the loss of electrons.
B. Oxidation is the loss of electrons, while reduction is the gain of electrons.
C. Oxidation and reduction both involve the loss of electrons.
D. Oxidation and reduction both involve the gain of electrons.

Answer:

QUESTION 217

What is the correct IUPAC name for the compound with the molecular formula C4H10?

    A.   Methane
    B.   Butane
    C.   Propane
    D.   Ethene

Answer:

QUESTION 218

Which factor affects the strength of an ionic bond?

    A.   Number of shared electrons
    B.   Electronegativity of the atoms involved
    C.   Atomic size of the atoms involved
    D.   Presence of lone pairs of electrons

Answer:

QUESTION 219

In a metallic bond, what primarily contributes to the ability of metals to conduct electricity?

    A.   The mobility of electrons in the metal lattice
    B.   The presence of covalent bonds between metal atoms
    C.   The transfer of electrons between metal atoms
    D.   The abundance of positively charged metal ions

Answer:

QUESTION 220

In a chemical equilibrium, what does it mean if Qc (reaction quotient) is greater than Kc (equilibrium constant)?

    A.   The reaction is not at equilibrium and will shift to the right to reach equilibrium.
    B.   The reaction is not at equilibrium and will shift to the left to reach equilibrium.
    C.   The reaction is at equilibrium.
    D.   It is impossible for Qc to be greater than Kc.

Answer:

QUESTION 221

In an aqueous solution with a hydroxide ion concentration of 1.0×10^{-5} M, what is the pOH value?

    A.   5
    B.   7
    C.   9
    D.   14

Answer:

QUESTION 222

For the reaction below, which species is undergoing reduction?

$Cu^{2+}(aq) + 2Ag(s) \rightarrow Cu(s) + 2Ag^+(aq)$

- A. $Cu^{2+}(aq)$
- B. $Ag(s)$
- C. $Cu(s)$
- D. $Ag^+(aq)$

Answer:

QUESTION 223

Which of the following statements is true regarding electrochemical cells?

- A. Electrolytic cells convert chemical energy into electrical energy.
- B. Electrochemical cells use electrical energy to drive non-spontaneous redox reactions.
- C. Electrochemical cells operate without any electrodes.
- D. Electrolytic cells convert electrical energy into chemical energy.

Answer:

QUESTION 224

Which of the following best describes an exothermic chemical reaction?

- A. It absorbs heat from the surroundings.
- B. It releases heat to the surroundings.
- C. It neither absorbs nor releases heat.
- D. It causes a change in temperature but does not involve heat.

Answer:

QUESTION 225

During a laboratory experiment, a student mixes two clear, colorless liquids together and observes the formation of a solid precipitate. Which type of chemical reaction is most likely to have occurred?

- A. Combustion reaction.
- B. Redox reaction.
- C. Acid-base reaction.
- D. Double replacement reaction.

Answer:

QUESTION 226

A chemistry teacher is conducting a reaction rate experiment and observes that the rate of the reaction is affected by the presence of a light source. The teacher concludes that the reaction is a photochemical reaction. What is the most likely outcome of this photochemical reaction?

- A. The reaction rate increases in the presence of light.
- B. The reaction rate decreases in the presence of light.
- C. The reaction rate remains the same in the presence of light.
- D. The reaction becomes spontaneous in the presence of light.

Answer:

QUESTION 227

The reaction between hydrochloric acid (HCl) and sodium hydroxide (NaOH) can be represented by the following balanced equation:

HCl + NaOH → NaCl + H2O

If 25 mL of 0.1 M HCl reacts with excess sodium hydroxide, what volume of 0.2 M NaOH solution is required to react completely with HCl?

A. 12.5 mL
B. 25 mL
C. 50 mL
D. 100 mL

Answer:

QUESTION 228

Which statement is true regarding an irreversible chemical reaction?

A. The reaction reaches equilibrium.
B. The products can react to reform the reactants.
C. The reaction rate can be altered by changing the concentration.
D. The reaction proceeds in one direction, and the products cannot convert back to the reactants.

Answer:

QUESTION 229

A chemistry teacher is investigating the effect of a catalyst on the rate of a chemical reaction. During the experiment, the teacher notices that the reaction rate increases initially but starts to decrease after a certain time. What could be a plausible explanation for this observation?

A. The catalyst is unstable and degrades over time.
B. The reaction has reached equilibrium, limiting further product formation.
C. The concentration of reactants is too high for the catalyst to be effective.
D. The temperature of the reaction mixture decreases during the experiment.

Answer:

QUESTION 230

A solution contains 20 grams of glucose (C6H12O6) dissolved in 100 grams of water. What is the mass percent of glucose in the solution?

A. 5%.
B. 10%.
C. 16.67%.
D. 20%.

Answer:

QUESTION 231

What is the correct definition of the mole in chemistry?

A. The number of atoms in 1 gram of an element.
B. The mass of 6.022×10^{23} atoms or molecules of a substance.
C. The volume occupied by 6.022×10^{23} atoms of a gas at STP.
D. The ratio of the atomic mass of an element to the mass of one mole of carbon-12.

Answer:

QUESTION 232

Which of the following applications of electrochemistry is used to protect iron and steel from corrosion?

    A. Galvanizing
    B. Electroplating
    C. Electrolysis
    D. Fuel cells

Answer:

QUESTION 233

What type of reaction is the rusting of iron (Fe) in the presence of oxygen (O2) and water (H2O)?

    A. Combination reaction.
    B. Decomposition reaction.
    C. Redox reaction.
    D. Single replacement reaction.

Answer:

QUESTION 234

A chemistry teacher is investigating the effect of particle size on the rate of a chemical reaction. Which of the following statements is correct?

    A. Smaller particle size increases the rate of reactions with higher activation energies.
    B. Larger particles increase the rate of exothermic reactions.
    C. Larger particle size increases the rate of diffusion-controlled reactions.
    D. Smaller particle size decreases the rate of surface area-dependent reactions.

Answer:

QUESTION 235

The reaction between hydrogen gas (H2) and oxygen gas (O2) to form water (H2O) can be represented by the balanced equation:

2H2 + O2 → 2H2O

What volume of O2, measured at STP (standard temperature and pressure), is required to completely react with 5 liters of H2, also measured at STP?

    A. 2.5 L
    B. 5 L
    C. 10 L
    D. 20 L

Answer:

QUESTION 236

Ms. Ramirez is planning a lesson on atomic structure for her chemistry class. She wants to sequence the learning activities to uncover common misconceptions and challenge her students to expand their understanding. Which of the following activities would be the most appropriate first step in her lesson plan?

    A. Giving a brief lecture on the history of atomic theory and the contributions of different scientists.
    B. Asking students to draw the Bohr model of an atom for a specific element.
    C. Providing students with a set of true/false statements about atomic structure and asking them to discuss their answers in pairs.
    D. Conducting a lab activity where students analyze the behavior of subatomic particles.

Answer:

QUESTION 237

Mr. Lee wants to assess his students' participation and understanding of a recent laboratory experiment on chemical reactions. He wants to gather both qualitative and quantitative data about their performance. Which assessment method is most appropriate for this purpose?

    A. Having students complete a self-assessment survey on their laboratory skills and understanding.
    B. Administering a standardized test on general chemistry concepts.
    C. Using a portfolio approach to collect students' laboratory reports and reflections.
    D. Assigning a group project to students and evaluating their teamwork during the experiment.

Answer:

QUESTION 238

When two solutions of different substances are mixed together, and a solid forms as a result, what type of reaction is likely to have occurred?

    A. Acid-base reaction
    B. Precipitation reaction
    C. Redox reaction
    D. Combustion reaction

Answer:

QUESTION 239

Ms. Thompson is teaching her chemistry class about the concept of stoichiometry. She wants her students to make systematic observations and measurements to understand stoichiometric calculations fully. Which of the following activities would be most effective in achieving her objective?

    A. Presenting a series of stoichiometry problems and working through the solutions on the board.
    B. Providing students with a set of chemical equations and asking them to balance the equations.
    C. Conducting a lab activity where students can measure the masses of reactants and products in a chemical reaction.
    D. Asking students to read a chapter in the textbook about stoichiometry.

Answer:

QUESTION 240

Mr. Johnson is teaching a chemistry class with diverse students, including English-language learners and students with special needs. He wants to ensure that all students can comprehend content-related texts effectively. Which instructional strategy would be most appropriate for achieving this goal?

    A. Encouraging collaboration among students to discuss the texts.
    B. Providing audio recordings of the texts for auditory learners.
    C. Using only traditional textbooks for consistent learning experiences.
    D. Assigning written summaries of the texts for assessment.

Answer:

QUESTION 241

Ms. Evans is introducing the concept of chemical reactions to her chemistry class. She wants to guide her students in making connections with their prior knowledge to understand the new topic better. Which of the following approaches would best help Ms. Evans achieve this goal?

    A. Showing a video demonstration of chemical reactions and asking students to write down their observations.
    B. Providing students with a list of chemical equations for them to memorize.
    C. Engaging students in a class discussion about the importance of chemical reactions in everyday life.
    D. Assigning a worksheet with a set of chemical reactions to balance.

Answer:

QUESTION 242

Mrs. Lee wants to plan a chemistry activity that encourages collaboration among her students. Which activity would be most suitable for fostering collaboration and group interaction?

A. Individual research project with no group involvement.
B. Competitive quiz with individual performance assessment.
C. Group experiment where students work together to achieve a common goal.
D. Listening to pre-recorded lectures with individual note-taking.

Answer:

QUESTION 243

Mr. Davis is teaching a unit on chemical reactions for his chemistry class. He wants his students to make systematic observations and measurements to understand these reactions fully. Which of the following activities would be most effective in achieving his objective?

A. Presenting a series of chemical reactions and asking students to identify the reactants and products.
B. Providing students with a list of chemical reaction equations for them to memorize.
C. Asking students to read a chapter in the textbook about chemical reactions and answer comprehension s.
D. Conducting a lab activity where students can measure the temperature changes during different chemical reactions.

Answer:

QUESTION 244

Mr. Rodriguez is teaching a chemistry class, and he wants to encourage students to develop their scientific s for an upcoming project. Which approach would best assist students in generating meaningful scientific s and hypotheses?

A. Providing a list of pre-defined s related to the project topic for students to choose from.
B. Assigning a fixed set of s and requiring students to answer them in the project.
C. Engaging students in brainstorming sessions and guiding them to formulate their s.
D. Encouraging students to use the internet as the primary resource for generation.

Answer:

QUESTION 245

Mr. Brown is teaching his chemistry class about the concept of chemical bonding. He wants his students to develop higher-level thinking skills and logical reasoning. Which of the following inquiry strategies would best achieve his objective?

A. Demonstrating the formation of different types of chemical bonds and explaining the concept to the students.
B. Giving students a set of chemical formulas and asking them to identify the type of chemical bond in each compound.
C. Asking students to read a chapter on chemical bonding in the textbook and answer comprehension s.
D. Providing students with a list of chemical bonds and asking them to memorize their properties.

Answer:

QUESTION 246

In a precipitation reaction, 200 mL of 0.2 M silver nitrate ($AgNO_3$) is mixed with 150 mL of 0.15 M sodium chloride ($NaCl$). What mass of silver chloride ($AgCl$) precipitate is formed?

A. 0.20 g
B. 0.24 g
C. 0.30 g
D. 0.36 g

Answer:

QUESTION 247

Which of the following molecules has a trigonal pyramidal shape?

A. CH4 (methane)
B. NH3 (ammonia)
C. H2O (water)
D. CO2 (carbon dioxide)

Answer:

QUESTION 248

Which of the following substances exhibits metallic bonding?

A. Sodium chloride (NaCl)
B. Iron (Fe)
C. Carbon dioxide (CO2)
D. Methane (CH4)

Answer:

QUESTION 249

A chemical reaction is conducted at constant pressure. If the volume of the system decreases during the reaction, what can be said about the enthalpy change (ΔH) and the heat (q) exchanged during the reaction?

A. ΔH is positive, and q is positive.
B. ΔH is negative, and q is positive.
C. ΔH is positive, and q is negative.
D. ΔH is negative, and q is negative.

Answer:

QUESTION 250

Ms. Patel is teaching her chemistry class about acids and bases. She wants her students to develop higher-level thinking skills and logical reasoning. Which of the following inquiry strategies would best achieve her objective?

A. Providing students with a list of acid-base properties and asking them to match them with the correct definitions.
B. Conducting a lab activity where students can test the pH of different substances and identify whether they are acids or bases.
C. Asking students to read a chapter on acids and bases in the textbook and answer comprehension s.
D. Demonstrating a neutralization reaction and explaining the concept to the students.

Answer:

QUESTION 251

A chemistry teacher wants to evaluate the reliability and validity of a new assessment instrument they plan to use in their class. Which action should the teacher take?

A. Administering the assessment to the students and grading it without considering the results.
B. Reviewing the assessment content with other teachers to get their opinions.
C. Comparing the results of the new assessment with results from a well-established assessment.
D. Asking the students for their feedback on how challenging the assessment was.

Answer:

QUESTION 252

A solution contains 1.5 moles of solute X dissolved in 500 mL of water. Another solution contains 2 moles of solute Y dissolved in 1 liter of water. Which solution has a higher concentration, and what is the molarity of that solution?

    A.   The solution with solute X has a higher concentration, and its molarity is 3 M.
    B.   The solution with solute X has a higher concentration, and its molarity is 2 M.
    C.   The solution with solute Y has a higher concentration, and its molarity is 2 M.
    D.   The solution with solute Y has a higher concentration, and its molarity is 3 M.

Answer:

QUESTION 253

Ms. Anderson is teaching her chemistry class about the concept of chemical reactions. She wants her students to develop higher-level thinking skills and scientific problem-solving abilities. Which of the following inquiry strategies would best achieve her goal?

    A.   Demonstrating a series of chemical reactions and asking students to identify the reactants and products.
    B.   Providing students with a set of chemical reaction equations and asking them to balance the equations.
    C.   Giving students a real-world problem that requires identifying a chemical reaction and proposing a solution based on their understanding.
    D.   Asking students to memorize a list of chemical reactions and their corresponding formulas.

Answer:

QUESTION 254

Which safety precaution is essential when working with corrosive chemicals?

    A.   Wearing gloves and lab coat.
    B.   Using glassware made of standard glass.
    C.   Storing chemicals in direct sunlight.
    D.   Mixing chemicals in an open environment.

Answer:

QUESTION 255

Which of the following best describes an element?

    A.   A substance composed of two or more chemically bonded atoms.
    B.   A substance that cannot be broken down into simpler substances by chemical means.
    C.   A mixture of two or more elements in fixed proportions.
    D.   A substance that undergoes a nuclear reaction to form new isotopes.

Answer:

QUESTION 256

A chemistry teacher is explaining the concept of molarity to her students. To illustrate the concept, she uses a glass of water and a glass of orange juice. She adds the same amount of sugar to both glasses and stirs until the sugar dissolves. The students observe that the sugar dissolves faster in the water than in the orange juice. How can the students explain this observation?

    A.   The sugar molecules have a higher solubility in water compared to orange juice.
    B.   The water molecules have a higher kinetic energy, leading to faster dissolution.
    C.   The orange juice is denser than water, hindering the sugar's movement and dissolution.
    D.   The sugar molecules experience stronger attractive forces in water than in orange juice.

Answer:

QUESTION 257

A chemistry class is investigating the factors that affect the rate of a chemical reaction. They experiment with the reaction between hydrochloric acid (HCl) and magnesium ribbon (Mg) to produce hydrogen gas (H2). The students test the reaction under three different conditions: Room temperature, in an ice bath, and in a hot water bath. After each trial, they measure and compare the amount of hydrogen gas produced. Which of the following factors is the most likely cause for the observed differences in reaction rates?

    A.   Concentration of hydrochloric acid (HCl).
    B.   Surface area of the magnesium ribbon (Mg).
    C.   Temperature of the reaction environment.
    D.   Presence of a catalyst.

Answer:

QUESTION 258

Which statement is true about the atomic number of an element?

    A.   It is equal to the total number of protons and neutrons in the nucleus.
    B.   It represents the mass of an atom relative to the mass of a carbon-12 atom.
    C.   It determines the chemical properties of an element.
    D.   It changes when an element forms chemical bonds.

Answer:

QUESTION 259

A chemical reaction between substances A and B results in the formation of two products, X and Y. Which statement is correct regarding the law of conservation of mass for this reaction?

    A.   The total mass of A and B will be equal to the total mass of X and Y.
    B.   The total mass of A and B will be less than the total mass of X and Y.
    C.   The total mass of A and B will be greater than the total mass of X and Y.
    D.   The total mass of A and B cannot be determined from the information provided.

Answer:

QUESTION 260

Which notation represents the electron configuration of an atom?

    A.   A series of letters representing the atom's subatomic particles (e.g., Xe for Xenon)
    B.   A diagram showing the arrangement of protons and neutrons in the atomic nucleus
    C.   A representation of the distribution of electrons in energy levels and sublevels
    D.   A grid-like structure showing the arrangement of elements in order of increasing atomic number

Answer:

QUESTION 261

Two chemistry students are investigating the acidity of different household substances. They have a solution of hydrochloric acid (pH 1) and a solution of acetic acid (vinegar, pH 3). They use pH paper to test the acidity of several substances, including lemon juice (pH 2) and ammonia solution (pH 11). The students wonder if a lower pH value always indicates a stronger acid and if a higher pH value always indicates a stronger base. How would you respond to their inquiry?

    A.   Yes, a lower pH value always indicates a stronger acid, and a higher pH value always indicates a stronger base.
    B.   No, the pH value alone does not determine the strength of an acid or base. Other factors must be considered.
    C.   Yes, a lower pH value indicates a stronger acid, but a higher pH value indicates a weaker base.
    D.   No, the pH value is irrelevant when determining the strength of an acid or a base.

Answer:

This page is intentionally left blank.

Chapter 2 – Answers and Explanations

QUESTION 1

Answer: C

Explanation: The nature of science in chemistry, like in any scientific field, is characterized by the dynamic nature of knowledge. Scientific theories are not considered absolute truths but rather our best explanations based on available evidence. As new evidence and data emerge, scientific theories may be revised or refined to better align with our understanding of the natural world.

QUESTION 2

Answer: B

Explanation: Encouraging critical thinking and scientific reasoning involves fostering an inclusive environment where diverse viewpoints are acknowledged and respected. The teacher should guide the students to recognize the difference between personal beliefs and scientific evidence. Emphasizing the importance of evidence-based arguments and providing opportunities for students to critically evaluate scientific literature and data will help them develop a deeper understanding of the topic and make informed conclusions based on evidence.

QUESTION 3

Answer: C

Explanation: Osmosis is the process of water molecules moving from a region of lower solute concentration (high water concentration) to a region of higher solute concentration (low water concentration) through a semipermeable membrane. The objective is to equalize the concentration of solute on both sides of the membrane. Option A's description is incorrect because osmosis involves the movement of water molecules, not solute molecules. Option B's description is not accurate because osmosis is specifically about water movement. Option D's statement is also inaccurate because osmosis is not related to pressure, making option C the correct answer.

QUESTION 4

Answer: A

Explanation: A catalyst is a substance that increases the rate of a chemical reaction by lowering its activation energy. It participates in the reaction but remains unchanged and is not consumed in the process.

QUESTION 5

Answer: B

Explanation: Promoting scientific literacy helps students develop the skills to assess, analyze, and interpret scientific information and claims. This empowers them to make informed decisions and understand the relevance of chemistry in their daily lives, regardless of their career paths.

QUESTION 6

Answer: A

Explanation: Titration is a technique used to determine the concentration of a solution by reacting it with a standard solution of known concentration until the reaction is complete.

QUESTION 7

Answer: D

Explanation: The student's observations suggest that increasing the concentration of one reactant results in a faster reaction rate. This is consistent with the collision theory, which states that higher reactant concentrations lead to a higher frequency of collisions between reactant molecules, thus increasing the rate of reaction. Therefore, increasing the concentration of the reactant leads to a faster reaction rate, making option D the correct answer.

QUESTION 8

Answer: B

Explanation: The dissolution of ammonium nitrate in water is an endothermic process. When ammonium nitrate dissolves in water, it absorbs heat from the surroundings, causing the temperature of the solution to decrease. Endothermic reactions require energy input to proceed, making option B the correct answer.

QUESTION 9

Answer: D

Explanation: In a polar covalent bond, the electrons are not shared equally between the atoms, resulting in a partial positive charge on one atom and a partial negative charge on the other. Water (H2O) is a classic example of a polar covalent compound, where oxygen exerts a stronger pull on the shared electrons compared to hydrogen, leading to a partial negative charge on oxygen and partial positive charges on the hydrogen atoms, making option D the correct answer.

QUESTION 10

Answer: D

Explanation: In a dynamic equilibrium, the forward and reverse reactions are occurring at the same rate, resulting in no net change in the concentration of reactants and products over time. Option D represents a reversible reaction reaching equilibrium, while the other options involve processes that do not reach a stable balance of reactants and products.

QUESTION 11

Answer: C

Explanation: A controlled experiment involves comparing different groups under controlled conditions. In this case, it would involve testing the three catalysts in identical reaction conditions to compare their effectiveness.

QUESTION 12

Answer: D

Explanation: Acids are substances that release hydrogen ions (H+) in water, lowering the pH of a solution. They have a sour taste, turn blue litmus paper red, and react with metals to produce hydrogen gas. However, acids do not increase the pH; rather, they decrease it, making option D the correct answer.

QUESTION 13

Answer: C

Explanation: The trend observed on the graph can be explained by the fact that at higher temperatures, the kinetic energy of water molecules increases. This increase in kinetic energy weakens the intermolecular forces between the solid substance and water molecules, making it easier for the solid to dissolve in the solvent.

QUESTION 14

Answer: D

Explanation: The observation that the reaction rate increased with an increase in temperature indicates a positive correlation between the two variables. This suggests that increasing temperature enhances the reaction rate, leading to the conclusion that the reaction is accelerated at higher temperatures.

QUESTION 15

Answer: C

Explanation: Allotropes are different forms of the same element in the same physical state. Diamond and graphite are two common allotropes of carbon. Diamond is a three-dimensional network of carbon atoms, whereas graphite is composed of two-dimensional layers of carbon atoms. Methane, ethanol, and sodium chloride are not allotropes of carbon, making option C the correct answer.

QUESTION 16

Answer: C

Explanation: The observation that the litmus paper turns blue indicates that the solution is basic (alkaline). However, it does not provide information about the exact pH value, concentration, or molarity of the solution. To determine the identity of the solute, further tests or additional information would be needed.

QUESTION 17

Answer: B

Explanation: Student B's method involved performing a series of chemical reactions with the unknown powder, which provided multiple observations of its behavior and properties. This approach is more reliable as it offers a more comprehensive understanding of the substance's chemical characteristics. While spectroscopy (Student A's method) can provide valuable information, relying on a single spectral analysis may lead to potential errors or misinterpretations. The confirmation from multiple chemical reactions makes Student B's method more reliable.

QUESTION 18

Answer: C

Explanation: In scientific research, unexpected results are not uncommon and can lead to new discoveries. To uphold the principles of scientific inquiry, the students should critically analyze the unexpected data, identify potential sources of error or bias, and consider alternative explanations for their observations. This process of revising their hypothesis based on the evidence will help them refine their understanding of the chemical compound and contribute to the advancement of knowledge in chemistry.

QUESTION 19

Explanation: The observation can be explained by the reactivity series of metals. Zinc is more reactive than copper and belongs to the group of transition metals, which tend to be more prone to chemical reactions. Copper, on the other hand, is less reactive and does not readily react with hydrochloric acid. The reactivity series of metals helps predict the order in which metals will react with acids and other substances.

QUESTION 20

Answer: A

Explanation: Technology in chemistry education can enhance the learning experience by offering virtual simulations that replicate real-life experiments. This approach allows students to explore different scenarios, collect data, and make observations in a safe and cost-effective manner, promoting a deeper understanding of chemical principles.

QUESTION 21

Answer: C

Explanation: Titration is a technique used to determine the concentration of a solution by reacting it with a solution of known concentration. The point at which the reaction is complete is indicated by a color change or other suitable indicator.

QUESTION 22

Answer: C

Explanation: Control groups are an integral part of experimental design in chemistry because they provide a baseline for comparison. In an experiment, the control group receives no treatment or the standard treatment, while other groups undergo specific manipulations. By comparing the results of the control group with those of the manipulated groups, researchers can assess the specific effects of the variables they are investigating. This allows them to isolate and attribute any observed changes or effects to the manipulated variables rather than other factors.

QUESTION 23

Answer: D

Explanation: Catalysts generally increase the rate of a chemical reaction by lowering the activation energy, making it easier for the reaction to occur. However, in some cases, other factors may be limiting the reaction rate. It is possible that the reaction is already proceeding at its maximum rate without the need for a catalyst, or there may be other factors such as temperature, concentration of reactants, or the nature of the reactants that are influencing the rate more significantly. It is essential to consider all possible factors that could affect the reaction rate to understand the unexpected result.

QUESTION 24

Answer: C

Explanation: Beta decay involves the transformation of a neutron into a proton, accompanied by the release of a beta particle (an electron). In this case, the atom has 7 protons, 7 neutrons, and 7 electrons before beta decay. During beta decay, one of the neutrons changes into a proton, resulting in an atom with 8 protons, 6 neutrons, and 8 electrons. The number of protons remains the same, the number of neutrons decreases by one, and the number of electrons increases by one to maintain charge neutrality.

QUESTION 25

Answer: C

Explanation: Using Charles's Law (the law of volumes), at constant pressure, the volume of a gas is directly proportional to its temperature in Kelvin. The initial temperature is 27°C, which is 300 Kelvin (273 + 27). The final temperature is 127°C, which is 400 Kelvin (273 + 127). Since the temperature has increased by a factor of 400/300 = 4/3, the volume will also increase by the same factor. Starting with a volume of 4 liters, the new volume will be 4 liters * 4/3 = 8 liters.

QUESTION 26

Answer: D

Explanation: Ionization energy refers to the energy required to remove an electron from an atom or ion. Elements with higher ionization energies have a stronger hold on their electrons. Among the given elements, silicon (Si) has the highest ionization energy. As you move across a period in the periodic table from left to right, the ionization energy generally increases, and silicon is farther to the right than sodium, magnesium, and aluminum.

QUESTION 27

Answer: B

Explanation: Specific heat capacity is the amount of heat energy required to raise the temperature of a substance by a certain amount. A higher specific heat capacity means the substance can absorb more heat energy for the same temperature change, resulting in a lower temperature change.

QUESTION 28

Answer: B

Explanation: Dissolving a solid solute in a solvent is an exothermic process because energy is released as the solute-solvent interactions form.

QUESTION 29

Answer: C

Explanation: Even though the reaction has a positive enthalpy change (endothermic), it can still occur spontaneously and rapidly if it is catalyzed. A catalyst lowers the activation energy, making it easier for the reaction to proceed, even though it requires energy input.

QUESTION 30

Answer: A

Explanation: Sodium (Na) is likely to have similar chemical properties and reactivity to other alkali metals like potassium (K) and lithium (Li) since they all belong to the same group (Group 1) on the periodic table. Elements in the same group have similar outer electron configurations, which determines their chemical behavior and reactivity.

QUESTION 31

Answer: C

Explanation: When a gas is compressed at constant temperature, its average kinetic energy remains constant. The temperature of the gas is directly proportional to its average kinetic energy, and since the temperature is constant in this scenario, the average kinetic energy does not change. However, the pressure and volume of the gas will change as a result of compression.

QUESTION 32

Answer: C

Explanation: Electrons in an atom are arranged in energy levels or shells. Electrons in higher energy levels are farther from the nucleus and have more energy than electrons in lower energy levels, which are closer to the nucleus and have lower energy.

QUESTION 33

Answer: A

Explanation: John Dalton proposed the first modern atomic theory in the early 19th century, introducing the idea of atoms as indivisible particles. Niels Bohr (b) described the planetary model of the atom, Dmitri Mendeleev (c) is known for developing the periodic table, and Ernest Rutherford (d) conducted the gold foil experiment that led to the discovery of the atomic nucleus.

QUESTION 34

Answer: C

Explanation: In a spontaneous process, the entropy (ΔS) of the system and the surroundings always increases, leading to an overall increase in the total entropy.

QUESTION 35

Answer: A

Explanation: The significant increase in potential energy for the products indicates that the reaction is endothermic and requires energy to proceed.

QUESTION 36

Answer: C

Explanation: When a molecule has a trigonalbipyramidal electron domain geometry and two lone pairs, the molecular geometry is T-shaped.

QUESTION 37

Answer: D

Explanation: The modern periodic table is arranged based on the atomic number of elements. Atomic number represents the number of protons in an atom's nucleus, and elements are arranged in ascending order of atomic number. This arrangement reflects the periodicity of properties and allows for the grouping of elements with similar chemical behavior into columns (groups) and periods (rows).

QUESTION 38

Answer: C

Explanation: The strong nuclear force is the interaction responsible for holding protons and neutrons together in the atomic nucleus. This force is much stronger than the electromagnetic force (which causes repulsion between protons) and overcomes the repulsive forces between positively charged protons, keeping the nucleus stable.

QUESTION 39

Answer: C

Explanation: The phase of a substance at a given temperature and pressure is determined primarily by the arrangement and interactions of its molecules or atoms. Different arrangements lead to different physical states, such as solid, liquid, or gas, depending on the strength of intermolecular forces and the thermal energy present in the substance.

QUESTION 40

Answer: D

Explanation: The 4d sublevel can hold a maximum of 10 electrons. The electron configuration of the 4d sublevel is 4d1 to 4d10, accommodating a total of 10 electrons.

QUESTION 41

Answer: B

Explanation: Gas pressure is caused by the constant collisions of gas particles with each other and the walls of their container, as stated by the kinetic molecular theory. Options (a), (c), and (d) do not explain the origin of gas pressure.

QUESTION 42

Answer: C

Explanation: Bomb calorimetry is a type of calorimetry that measures the heat exchanged at constant volume. In this setup, the reactants are placed in a sealed container (the bomb), and the reaction occurs at a constant volume, allowing for accurate measurements of the heat released or absorbed during the process.

QUESTION 43

Answer: B

Explanation: Exothermic reactions release heat to the surroundings, resulting in a decrease in the enthalpy of the system.

QUESTION 44

Answer: A

Explanation: According to IUPAC rules, the compound with the molecular formula CH3COOH is named acetic acid.

QUESTION 45

Answer: B

Explanation: Noble gases, also known as Group 18 elements, have a complete outer electron shell (octet) with eight valence electrons, except for helium, which has only two. This stable electron configuration makes them chemically inert and unreactive under normal conditions, hence their name "noble gases."

QUESTION 46

Answer: D

Explanation: The average velocity of gas molecules is inversely proportional to the square root of their molar masses, according to Graham's law of effusion and diffusion. Since hydrogen (H2) has the lowest molar mass among the given options, it will have the highest average velocity at the same temperature.

QUESTION 47

Answer: C

Explanation: Isotopes of the same element have the same number of protons since the number of protons defines the element. However, they have different numbers of neutrons, resulting in variations in their mass numbers. Because of the difference in neutrons, isotopes may have slightly different physical properties but exhibit similar chemical properties due to the same number of protons and electrons.

QUESTION 48

Answer: A

Explanation: The units for the gas constant R in the ideal gas law are L·atm/(mol·K). Option (b) represents density, option (c) is molarity, and option (d) is molar heat capacity.

QUESTION 49

Answer: B

Explanation: By not accounting for the heat gained or lost by the calorimeter itself, the student would underestimate the amount of heat absorbed by the water, leading to an erroneously low value for the calculated heat capacity of the unknown metal.

QUESTION 50

Answer: C

Explanation: Evaporation of a liquid leads to an increase in entropy as the molecules disperse and move from a more ordered liquid phase to a more dispersed vapor phase.

QUESTION 51

Answer: C

Explanation: When a molecule has a tetrahedral electron domain geometry and no lone pairs, the molecular geometry is also tetrahedral.

QUESTION 52

Answer: A

Explanation: Silicon (Si) is a metalloid, which is an element that has properties intermediate between those of metals and nonmetals. Metalloids are found along the "stair-step" line on the periodic table. Silicon exhibits both metallic and nonmetallic properties, making it an essential material in semiconductor technology and other industrial applications.

QUESTION 53

Answer: B

Explanation: The kinetic molecular theory explains the compressibility of gases. According to the theory, gases consist of particles in constant, random motion, and they have significant amounts of empty space between them. When pressure is applied to a gas, the gas particles can be compressed closer together, resulting in a decrease in volume.

QUESTION 54

Answer: B

Explanation: The net charge of an atom is determined by comparing the number of protons and electrons. In this case, the atom has 15 protons (positive charge) and 16 electrons (negative charge). The net charge is calculated as 15 (protons) - 16 (electrons) = -1.

QUESTION 55

Answer: B

Explanation: The Second Law of Thermodynamics states that in any spontaneous process, the total entropy of an isolated system always increases over time, indicating an increase in disorder.

QUESTION 56

Answer: D

Explanation: A positive ΔH and negative ΔS both indicate a non-spontaneous process. Without further information, the solubility of the solid cannot be determined solely based on these thermodynamic parameters.

QUESTION 57

Answer: B

Explanation: Adiabatic expansion involves the gas doing work on its surroundings, which results in the release of energy, making it an exothermic process.

QUESTION 58

Answer: D

Explanation: Optical isomers (enantiomers) have the same molecular formula and the same arrangement of atoms but are mirror images of each other and differ in their interaction with plane-polarized light.

QUESTION 59

Answer: A

Explanation: An aldehyde contains the aldehyde functional group, which is characterized by the presence of a carbonyl group (C=O) at the end of a carbon chain.

QUESTION 60

Answer: C

Explanation: Molecule B, with its bent structure, is likely to exhibit hydrogen bonding if it contains hydrogen atoms bonded to electronegative atoms with lone pairs. Hydrogen bonding is stronger than dipole-dipole interactions and London dispersion forces, which could be present in both molecules due to their polarity. Ion-dipole interactions are not relevant here since there are no ions involved.

QUESTION 61

Answer: D

Explanation:

A hydrogen bond is a type of intermolecular bond formed between a partially positive hydrogen atom (from one molecule) and a highly electronegative atom (such as oxygen, nitrogen, or fluorine) in another molecule.

QUESTION 62

Answer: B

Explanation: For an exothermic reaction, increasing the temperature will shift the equilibrium position towards the reactants (Le Châtelier's principle). As a result, the concentration of products will decrease, leading to a decrease in the value of the equilibrium constant (Kc).

QUESTION 63

Answer: C

Explanation: A buffer solution is a solution that can resist changes in pH when small amounts of acid or base are added to it. A solution of a weak acid and its conjugate base is an example of a buffer. When a small amount of acid is added, the conjugate base can react with it, and when a small amount of base is added, the weak acid can neutralize it, keeping the pH relatively stable.

QUESTION 64

Answer: D

Explanation: The voltage output (emf) of a galvanic cell is determined by the difference in standard reduction potentials between the two half-reactions. Using electrodes with higher standard reduction potentials will result in a larger voltage output, leading to a more powerful galvanic cell.

QUESTION 65

Answer: A

Explanation: Covalent bonds involve the sharing of electrons between two non-metal atoms, forming a strong bond that holds the atoms together in a molecule.

QUESTION 66

Answer: A

Explanation: The fact that Compound X conducts electricity when dissolved in an ionic liquid indicates that it must be an ionic compound. Ionic compounds do not conduct electricity in their solid state but become conductive when dissolved in an ionic liquid due to the dissociation of ions in solution.

QUESTION 67

Answer: A

Explanation:

The number of shared electrons does not directly affect the bond strength in a covalent bond. The bond strength primarily depends on the distance between the bonded atoms and the electronegativity difference between them.

QUESTION 68

Answer: B

Explanation: The equilibrium constant (Kc) is a constant value for a given reversible chemical reaction at a specific temperature. It does not depend on the initial concentrations of reactants and products but only on their concentrations at equilibrium. Kc is the same for the forward and reverse reactions since it represents the ratio of the concentrations of products to reactants at equilibrium.

QUESTION 69

Answer: A

Explanation: The sum of pH and pOH in a solution at 25°C is always equal to 14. This relationship is a result of the self-ionization of water, where the concentration of H^+ ions and OH^- ions in pure water is 1.0×10^{-7} M each, and their product (pH × pOH) is equal to 1×10^{-14}. Therefore, pH + pOH = $-\log(1 \times 10^{-7}) + (-\log(1 \times 10^{-7})) = 7 + 7 = 14$.

QUESTION 70

Answer: C

Explanation: The Nernst equation is used to calculate the cell potential of an electrochemical cell at non-standard conditions. It relates the cell potential (Ecell) to the standard cell potential (E°cell) and the reaction quotient (Q) using the concentrations of the species involved in the redox reaction.

QUESTION 71

Answer: D

Explanation: An endothermic chemical reaction is one that requires an input of energy from the surroundings to proceed. During the reaction, energy is absorbed, which often leads to a decrease in the surrounding temperature.

QUESTION 72

Answer: B

Explanation: London dispersion forces are the weakest among intermolecular forces. They arise from temporary fluctuations in electron distribution, and while individually weak, they can collectively influence the properties of gases at room temperature.

QUESTION 73

Answer: B

Explanation: Ionic compounds can conduct electricity both in their solid state (as ions are fixed in a lattice and can conduct when an electric field is applied) and when dissolved in water (as the ions dissociate and move freely in solution). Ionic bonding involves the transfer of electrons between elements with significantly different electronegativities, resulting in the formation of charged ions.

QUESTION 74

Answer: A

Explanation:

Bond enthalpy is the amount of energy required to break a specific bond in a molecule. Stronger bonds have higher bond enthalpy values since more energy is needed to overcome the attractive forces between the bonded atoms.

QUESTION 75

Answer: B

Explanation: A large value of the equilibrium constant (Kc) indicates that at equilibrium, the concentration of products is significantly higher than the concentration of reactants. In this case, Kc = 500, suggesting that the equilibrium favors the products.

QUESTION 76

Answer: C

Explanation: In an acid-base titration, the titration is complete when the amount of titrant (known solution of base or acid) added is stoichiometrically equivalent to the amount of the analyte solution (the solution of unknown concentration). This point is called the equivalence point, and it is reached when the moles of acid and base are in the ratio required for complete neutralization. At this point, the volume of the titrant added is equal to the volume of the analyte solution.

QUESTION 77

Answer: D

Explanation: A fuel cell is an electrochemical device that converts the chemical energy of a fuel (e.g., hydrogen) and an oxidizing agent (e.g., oxygen) into electrical energy through redox reactions. It does not involve the combustion of fuels like internal combustion engines.

QUESTION 78

Answer: D

Explanation: The rate of a chemical reaction is influenced by the concentration of the reactants. Higher concentrations of reactants generally lead to an increased rate of reaction because there are more particles available for collisions, resulting in more frequent successful collisions and faster reaction rates.

QUESTION 79

Answer: D

Explanation: The strength of hydrogen bonding between molecules is influenced by the presence of lone pairs of electrons in the molecules, which can participate in forming hydrogen bonds with other molecules.

QUESTION 80

Answer: A

Explanation:

The strength of a chemical bond depends on the distance between the nuclei of the bonded atoms. As the nuclei get closer together, the attraction between the positively charged nuclei and the shared electrons increases, resulting in a stronger bond.

QUESTION 81

Answer: C

Explanation: Catalysts do not affect the position of a chemical equilibrium. They only speed up the rate at which equilibrium is reached by providing an alternative pathway with a lower activation energy. Once equilibrium is reached, the position remains unchanged.

QUESTION 82

Answer: A

Explanation: Increasing the concentration of NOCl will disturb the equilibrium, according to Le Châtelier's principle, the system will shift in the direction that reduces the concentration of NOCl. In this case, it will shift towards the reactants (NO and Cl_2) to reduce the excess NOCl concentration.

QUESTION 83

Answer: A

Explanation: The pOH is calculated as the negative logarithm (base 10) of the hydroxide ion concentration in Molarity (M). pOH = -log(OH^-). To calculate pOH from pH, we can use the relationship: pH + pOH = 14. If the pH is 9, then pOH = 14 - 9 = 5.

QUESTION 84

Answer: B

Explanation: In a galvanic cell, the spontaneous generation of an electric current is due to the flow of electrons through the external circuit from the anode to the cathode. This electron flow constitutes the electric current, and it is the basis of the cell's ability to do work.

QUESTION 85

Answer: A

Explanation: Covalent bonds are the strongest type of chemical bond, as they involve the sharing of electrons between atoms, creating a strong bond that holds the atoms together in a molecule.

QUESTION 86

Answer: B

Explanation:

Breaking an ionic bond involves overcoming the electrostatic attraction between the oppositely charged ions, which requires energy input. Conversely, when ions come together to form an ionic bond, energy is released due to the attractive forces between the ions.

QUESTION 87

Answer: A

Explanation: Increasing the concentration of N2O4 will disturb the equilibrium, according to Le Châtelier's principle, the system will shift in the direction that reduces the concentration of N2O4, which means it will shift towards N2O4.

QUESTION 88

Answer: B

Explanation: Adding heat will shift the equilibrium position towards the endothermic direction, which is the formation of NO2. To restore the original equilibrium position, you need to shift it back towards N2O4. According to Le Châtelier's principle, decreasing the volume of the container will increase the pressure and favor the side with fewer moles of gas, which means it will shift towards N2O4.

QUESTION 89

Answer: B

Explanation: In an acid-base titration of a strong acid with a strong base, the sharp rise in pH to pH 7 indicates that the equivalence point has been reached. At the equivalence point, the moles of acid and base are in stoichiometrically equivalent amounts, resulting in a neutral solution (pH 7). Any further addition of the base beyond the equivalence point would lead to a basic solution with a higher pH.

QUESTION 90

Answer: A

Explanation: According to the Nernst equation, an increase in the concentration of reactants (Cu^{2+} ions in this case) in the half-cell undergoing reduction (the right half-cell) will lead to an increase in the cell potential. This is because an increase in reactant concentration drives the reaction towards completion, increasing the electron flow and resulting in a higher cell potential.

QUESTION 91

Answer: A

Explanation: In this setup, the copper electrode, being connected to the negative terminal of the external circuit, is the anode. It undergoes oxidation, losing electrons to form copper ions in the solution. Meanwhile, the silver electrode, connected to the positive terminal of the external circuit, is the cathode. It undergoes reduction, gaining electrons from the external circuit to form solid silver. The copper electrode gains mass due to the deposition of copper ions from the solution, while the silver electrode loses mass due to the formation of silver ions in the solution.

QUESTION 92

Answer: C

Explanation: In a polar covalent bond, one atom has a higher electronegativity than the other, resulting in a partial negative charge on the more electronegative atom and a partial positive charge on the less electronegative atom.

QUESTION 93

Answer: D

Explanation: A catalyst increases the rate of both the forward and reverse reactions equally. As a result, the equilibrium position remains unchanged, and the value of the equilibrium constant (Kc) remains the same.

QUESTION 94

Answer: C

Explanation: Bases are substances that, when dissolved in water, increase the concentration of hydroxide ions (OH^-). This behavior is what characterizes a base. According to the Brønsted-Lowry definition of bases, a base is a proton acceptor.

QUESTION 95

Answer: B

Explanation: The reducing agent is the species that undergoes oxidation and causes another species to be reduced. In this reaction, Cl_2 is the reducing agent because it causes K to undergo oxidation (from 0 to +1) while itself getting reduced (from 0 to -1).

QUESTION 96

Answer: C

Explanation: Chlorine gas (Cl_2) is produced during the electrolysis of sodium chloride (NaCl) when chloride ions (Cl-) are oxidized at the anode. Platinum (Pt) electrodes are commonly used for this purpose, as they are inert and do not react with the chlorine gas or the products of the reaction, making them suitable for the electrolysis of chloride ions.

QUESTION 97

Answer: C

Explanation: Chemical reactions involve the breaking of bonds in the reactants, which requires an input of energy (endothermic process). Conversely, when new bonds are formed in the products, energy is released (exothermic process). The overall energy change during a reaction depends on the difference between the energy absorbed and released in the process.

QUESTION 98

Answer: A

Explanation: A catalyst provides an alternative reaction pathway that requires less energy to overcome the activation barrier. By doing so, it lowers the activation energy, making it easier for the reaction to proceed and increasing the reaction rate.

QUESTION 99

Answer: B

Explanation: Catalysts work by providing an alternative reaction pathway with a lower activation energy. This lowered activation energy allows reactant molecules to more easily convert into products, effectively increasing the reaction rate. The catalyst itself is not consumed during the reaction and does not alter the overall energy change in the reaction; it simply accelerates the rate at which the reaction proceeds.

QUESTION 100

Answer: C

Explanation: The pressure applied to the system does not significantly affect the rate of dissolution of a solid solute in a liquid solvent. It is mainly influenced by particle size, temperature, and polarity.

QUESTION 101

Answer: A

Explanation: To determine the limiting reactant, the balanced equation is necessary. The balanced equation for the reaction is $C_2H_6 + 7O_2 \rightarrow 2CO_2 + 3H_2O$. Given that 1 mole of C_2H_6 requires 7 moles of O_2, if an equal number of moles of C_2H_6 and O_2 are used, the oxygen (O_2) will be in excess, making ethane (C_2H_6) the limiting reactant.

QUESTION 102

Answer: C

Explanation: Metallic bonding involves the sharing of delocalized electrons among a lattice of metal atoms. This "sea of electrons" allows metals to conduct electricity efficiently, making option c) the correct answer.

QUESTION 103

Answer: A

Explanation: To find the limiting reactant, we need to compare the mole ratios of reactants and products. The balanced chemical equation is 2A + 3B -> C + D. If 2 moles of A react, it requires (2 moles A * 3 moles B) / 2 moles A = 3 moles of B. Since we only have 3 moles of B, it is the limiting reactant. After the reaction, 1 mole of B remains (3 moles B - 2 moles B used in the reaction). And for A, 2 moles react, so 2 - 2 = 0 moles of A remains.

QUESTION 104

Answer: D

Explanation: Option D aligns with the strategy of assisting students in learning to identify potential sources of error in an inquiry-based scientific investigation. By guiding students to reflect on their experiments after completion, they can critically analyze their methods, observations, and results to identify any sources of error. This approach fosters a deeper understanding of experimental design and the importance of recognizing and mitigating potential biases in scientific investigations.

QUESTION 105

Answer: C

Explanation: In a double displacement reaction, two compounds exchange their ions to form new compounds. When two clear colorless solutions combine and a solid precipitate is formed, it indicates that a chemical reaction occurred between the two solutions, resulting in the formation of an insoluble product. This aligns with the characteristics of a double displacement reaction.

QUESTION 106

Answer: A

Explanation: At high temperatures, the catalyst may undergo deactivation due to structural changes or thermal degradation, leading to a decrease in its catalytic activity. As a result, the reaction rate starts to decrease despite the normally expected increase in reaction rate with higher temperatures.

QUESTION 107

Answer: D

Explanation: The Tyndall effect is the scattering of light by colloidal particles and larger particles in suspensions. It is not observed in solutions because the particles in a true solution are too small to scatter light.

QUESTION 108

Answer: B

Explanation: To determine the volume of 0.5 M hydrochloric acid (HCl) required for the complete reaction with 25 grams of calcium hydroxide ($Ca(OH)_2$), we start by calculating the number of moles of $Ca(OH)_2$. The molar mass of $Ca(OH)_2$ is 74.09 g/mol, so 25 grams corresponds to approximately 0.3375 moles of $Ca(OH)_2$. The balanced chemical equation for the reaction shows that one mole of $Ca(OH)_2$ reacts with two moles of HCl. Thus, to find the moles of HCl needed, we multiply the moles of $Ca(OH)_2$ by 2, yielding 0.675 moles of HCl. Using the formula for concentration, moles divided by molarity, we find the volume in liters, which is 1.35 L. To express this volume in milliliters (as per the answer choices), we multiply by 1000, giving us 1350 mL. However, since the answer choices are in tens, we divide this by 10, arriving at the closest choice of 140 mL. Therefore, 140 mL of 0.5 M HCl is required for the complete reaction with 25 grams of $Ca(OH)_2$.

QUESTION 109

Answer: C

Explanation: Ionic bonds involve the transfer of electrons from one atom to another, leading to the formation of oppositely charged ions that are attracted to each other, creating a bond.

QUESTION 110

Answer: B

Explanation: Percent yield is calculated by dividing the actual yield by the theoretical yield and then multiplying by 100%. Using the given percent yield of 80%, we can set up the equation: 80% = (Actual Yield / Theoretical Yield) * 100%. Let the actual yield be "Y" grams. Therefore, 80 = (Y / 50) * 100. Solving for Y, we get Y = (80 * 50) / 100 = 40 grams. Hence, the actual yield obtained by the student was 40 grams.

QUESTION 111

Answer: B

Explanation: Option B involves hands-on activities that require students to make systematic observations and measurements of different substances. By doing so, they actively engage with the concepts and develop a deeper understanding of the properties of matter.

QUESTION 112

Answer: D

Explanation: The teacher is investigating how temperature affects the rate of the chemical reaction while keeping the concentration of reactants constant. According to collision theory, increasing the temperature increases the kinetic energy of particles, leading to more frequent and energetic collisions between reactant molecules. As a result, the rate of the reaction increases.

QUESTION 113

Answer: A

Explanation: When substance C acts as a catalyst, it provides an alternate reaction pathway with a lower activation energy, increasing the reaction rate. The catalyst remains unchanged at the end of the reaction, explaining why the rate remains constant regardless of the concentrations of A, B, and C.

QUESTION 114

Answer: B

Explanation: In a balanced chemical equation, the number of atoms of each element should be the same on both sides of the equation. Option B has a balanced equation with one carbon (C) atom, four hydrogen (H) atoms, and four oxygen (O) atoms on both sides.

QUESTION 115

Answer: C

Explanation: The empirical formula indicates the simplest whole-number ratio of elements in a compound, which in this case is CH_2O. To find the molecular formula, we need to know the molar mass of the compound. The molar mass of 180 g/mol indicates that the molecular formula must be a multiple of the empirical formula. Option c) $C_6H_{12}O_6$ is the correct choice, as it has a molar mass of 180 g/mol, and it is the multiple of the empirical formula (CH_2O) by a factor of 6.

QUESTION 116

Answer: C

Explanation: A higher electronegativity difference between two bonded atoms indicates a stronger ionic bond. In such cases, one atom strongly attracts electrons, leading to the formation of ions with opposite charges that are held together by electrostatic forces.

QUESTION 117

Answer: A

Explanation: Exothermic reactions release energy in the form of heat to the surroundings. When the reaction is performed in a calorimeter, the surroundings (calorimeter and its contents) will absorb this released heat, causing an increase in temperature. Therefore, the temperature of the surroundings will increase.

QUESTION 118

Answer: C

Explanation: Option C facilitates building upon students' prior knowledge by discussing common household chemical reactions. This approach helps students relate the new concept of chemical equations to their everyday experiences, making it more meaningful and memorable.

QUESTION 119

Answer: B

Explanation: In a zero-order reaction, the reaction rate is independent of the concentration of the catalyst or reactants. The presence of a catalyst may speed up or slow down reactions, but in this case, the reaction follows a zero-order kinetics, meaning the rate is not affected by changes in the catalyst's concentration.

QUESTION 120

Answer: B

Explanation: A solution is a homogeneous mixture where one substance (the solute) is evenly distributed within another substance (the solvent) at the molecular level.

QUESTION 121

Answer: A

Explanation: To determine the limiting reactant, we need to calculate the moles of each reactant and compare them based on their stoichiometric ratio in the balanced equation. The balanced equation for the reaction is 2Mg + O2 → 2MgO. The moles of Mg and O2 can be calculated from their given masses, and then we can compare the mole ratio with the stoichiometric ratio to identify the limiting reactant.

QUESTION 122

Answer: D

Explanation: Similar to the previous , we find the empirical formula by converting the percentages into grams and then moles. The empirical formula of this compound is CH2O. To find the molecular formula, we divide the molar mass of the compound by the molar mass of the empirical formula. In this case, 120 g/mol ÷ 30 g/mol (molar mass of CH2O) gives 4. This means the molecular formula is four times the empirical formula: $C_4H_8O_4$. However, we want the lowest whole-number ratio, so we divide all the subscripts by 2, giving us $C_2H_4O_2$. Option d) $C_9H_{18}O_9$ is the correct choice.

QUESTION 123

Answer: C

Explanation: The strength and stability of a covalent bond are determined by the extent of overlap between the atomic orbitals of the bonding atoms. A stronger and more extensive overlap results in a stronger and more stable covalent bond.

QUESTION 124

Answer: C

Explanation: A buffer solution is a solution that resists changes in pH when small amounts of acid or base are added to it. Buffers are essential in various chemical laboratory and industrial processes where a stable pH environment is required. They consist of a weak acid and its conjugate base (or a weak base and its conjugate acid), which react with added acidic or basic species to minimize the change in pH.

QUESTION 125

Answer: D

Explanation: Option D is the most appropriate final step as it challenges students to apply their knowledge of chemical bonding to solve problems related to Lewis structures. It requires a deeper level of understanding and critical thinking, making it an effective concluding activity.

QUESTION 126

Answer: A

Explanation: The rate expression of a chemical reaction represents how the rate of the reaction depends on the concentrations of the reactants. In the given reaction, the rate is proportional to the concentrations of both reactants, A and B, based on the stoichiometric coefficients in the balanced equation.

QUESTION 127

Answer: B

Explanation: Henry's law states that the solubility of a gas in a liquid is directly proportional to the partial pressure of the gas above the liquid. Therefore, increasing the pressure will increase the solubility of the gas.

QUESTION 128

Answer: B

Explanation: Percent yield is the actual yield divided by the theoretical yield, multiplied by 100. First, calculate the theoretical yield of CO2 and H2O from the given mass of CH4. Then, find the actual yield by adding the masses of CO2 and H2O produced. Finally, calculate the percent yield using the formula mentioned earlier.

QUESTION 129

Answer: A

Explanation: To calculate the percent composition of nitrogen in each compound, we divide the molar mass of nitrogen by the molar mass of the whole compound and multiply by 100. Comparing the compounds, we find that ammonium nitrate (NH_4NO_3) has the highest percentage of nitrogen (approximately 35%).

QUESTION 130

Answer: C

Explanation: The bond length in a covalent bond is determined by the size of the bonded atoms and the strength of the bond. Generally, a stronger bond has a shorter bond length, and a weaker bond has a longer bond length.

QUESTION 131

Answer: A

Explanation: Molality (m) is defined as moles of solute per kilogram of solvent. First, we need to convert the mass of KCl to moles: Moles of KCl = 30 g / molar mass of KCl = 30 g / 74.55 g/mol ≈ 0.402 mol. Next, we convert the mass of water to kilograms: 500 g of water = 500 g / 1000 g/kg = 0.5 kg. Finally, we calculate the molality: Molality (m) = Moles of KCl / Kilograms of water = 0.402 mol / 0.5 kg ≈ 0.804 mol/kg. Therefore, the molality of the solution is approximately 0.804 mol/kg, which can be rounded to 0.8 mol/kg.

QUESTION 132

Answer: C

Explanation: Option C is the most appropriate first step as it engages students in a hands-on experience to observe the properties of different elements. By doing so, students can develop a foundation of knowledge about elements before diving into more complex topics like the organization of the periodic table.

QUESTION 133

Answer: C

Explanation: A lower activation energy means that the reactant molecules in reaction P can overcome the energy barrier more easily compared to reaction Q. As a result, reaction P will have a faster rate because it requires less energy for the reactants to transform into products.

QUESTION 134

Answer: C

Explanation: Colligative properties, such as boiling point elevation and freezing point depression, depend on the total number of solute particles in a solution, not their nature.

QUESTION 135

Answer: B

Explanation: An exothermic process releases heat energy to the surroundings. In option B, the reaction involves the decomposition of calcium carbonate (CaCO3) into calcium oxide (CaO) and carbon dioxide (CO2), and it releases heat during the process.

QUESTION 136

Answer: C

Explanation: To find the empirical formula, we need to determine the number of moles of carbon and hydrogen in the sample. First, we find the number of moles of CO_2 produced (1.56 grams) and calculate the number of moles of carbon in the sample. Then, we find the number of moles of H_2O produced (0.45 grams) and calculate the number of moles of hydrogen in the sample. The mole ratio of carbon to hydrogen is 2:2, which simplifies to 1:1. Thus, the empirical formula is $C_2H_2O_2$, which is option c).

QUESTION 137

Answer: C

Explanation: Stoichiometry is the study of the quantitative relationships between the amounts of reactants and products in a chemical reaction. It allows us to understand how much of each reactant is required and how much product can be formed based on the balanced chemical equation.

QUESTION 138

Answer: B

Explanation: To respect student diversity and make the lesson inclusive, Mr. Thompson should incorporate a variety of instructional materials that cater to different learning preferences. By using a mix of visuals, hands-on activities, auditory resources, and texts, he can meet the interests, abilities, and experiences of all students, including English-language learners and students with special needs. This approach ensures that each student has opportunities to engage with the content in a way that suits their learning style.

QUESTION 139

Answer: A

Explanation: Option A is the most appropriate final step as it challenges students to apply their knowledge of the states of matter to solve problems related to particle behavior. This activity requires critical thinking and the application of their understanding.

QUESTION 140

Answer: C

Explanation: In a first-order reaction (R), doubling the concentration of reactants will double the reaction rate since the rate is directly proportional to the concentration of one reactant. In contrast, in a second-order reaction (S), doubling the concentration of reactants will quadruple the reaction rate, as the rate is proportional to the square of the concentration of one reactant.

QUESTION 141

Answer: A

Explanation: Molarity is defined as moles of solute per liter of solution and is a common unit used to express the concentration of a solution.

QUESTION 142

Answer: B

Explanation: A covalent bond is formed when two atoms share a pair of electrons, allowing them to complete their outer electron shells and become more stable. This type of bond is commonly found in molecules like water (H2O) and methane (CH4).

QUESTION 143

Answer: B

Explanation: Mr. Smith's objective is to assess his students' understanding of the concepts learned in the unit on chemical reactions. The most suitable assessment method for this purpose is a formal written exam with multiple-choice and short-answer s. This method allows him to test students' theoretical knowledge, recall of key concepts, and problem-solving abilities related to chemical reactions. It aligns with the need to monitor and assess students' understanding of science concepts on an ongoing basis using appropriate assessment methods.

QUESTION 144

This question is intentionally removed.

QUESTION 145

Answer: A

Explanation: A net ionic equation shows only the species that participate directly in the reaction, excluding spectator ions. In option A, Ag^+ and Cl^- ions combine to form the insoluble $AgCl$ precipitate. The Na^+ and NO_3^- ions remain unchanged and are spectator ions.

QUESTION 146

Answer: C

Explanation: The empirical formula for this compound is CH_2O. To find the molecular formula, we divide the molar mass of the compound by the molar mass of the empirical formula (12 + 1 + 16). The result is approximately 4.29, which means the molecular formula is 4.29 times the empirical formula. Rounding to the nearest whole numbers, we get $C_6H_6O_3$, which is option c).

QUESTION 147

Answer: B

Explanation: To determine the mass of potassium sulfate ($K2SO4$) formed when 20 grams of potassium hydroxide (KOH) reacts with excess sulfuric acid ($H2SO4$), we start by utilizing the balanced chemical equation for the reaction: $2KOH + H2SO4 \rightarrow K2SO4 + 2H2O$. We initially calculate the moles of KOH by dividing its mass by its molar mass, resulting in approximately 0.3566 moles of KOH. Next, we use the mole ratio from the balanced equation to find the moles of K2SO4 produced for every mole of KOH. According to the equation, 2 moles of KOH produce 1 mole of K2SO4, yielding 0.1783 moles of K2SO4. Finally, to find the mass of K2SO4, we multiply the moles of K2SO4 by its molar mass, which is approximately 174.26 g/mol, resulting in an approximate mass of 31.10 grams of K2SO4 formed in the reaction. Therefore, when 20 grams of KOH reacts with excess H2SO4, about 31.10 grams of K2SO4 are produced.

QUESTION 148

Answer: C

Explanation: In the field of chemistry, it is crucial for students to develop a scientific attitude that includes skepticism and critical thinking. Rather than accepting claims at face value, students should be encouraged to and challenge scientific ideas, hypotheses, and theories. Emphasizing skepticism helps foster a deeper understanding of scientific concepts and promotes the development of analytical skills necessary for scientific inquiry.

QUESTION 149

Answer: A

Explanation: The observations indicate that the reaction of metal A with the acid was the most vigorous, producing a large amount of gas bubbles. This suggests that metal A is highly reactive with the acid, resulting in a rapid production of gas (likely hydrogen gas) during the reaction. The reactivity of metals with acids is related to their position in the reactivity series. More reactive metals, such as metal A, react more vigorously and readily with acids, leading to the production of a greater amount of gas bubbles, making option A the correct answer.

QUESTION 150

Answer: C

Explanation: Specific heat capacity (often denoted as "C") is the amount of heat energy required to raise the temperature of one gram of a substance by one degree Celsius. It is a property that depends on the type of substance.

QUESTION 151

Answer: B

Explanation: Option B illustrates the practical application of chemistry in assessing environmental pollution, which is a real-world problem. Analyzing soil samples involves chemical techniques and methods to detect and quantify pollutants, showcasing the relevance of chemistry in addressing environmental issues.

QUESTION 152

Answer: D

Explanation: The uncertainty value provided by the instrument represents the precision of the measurement. It should be included in the recorded value to express the level of uncertainty.

QUESTION 153

Answer: C

Explanation: Uranium-238 is a radioactive material. It undergoes radioactive decay, emitting radiation in the form of alpha particles, beta particles, and gamma rays.

QUESTION 154

Answer: A

Explanation: A decomposition reaction is a type of chemical reaction where a single compound breaks down into two or more simpler substances. In Reaction 1, hydrogen peroxide (H_2O_2) decomposes into water (H_2O) and oxygen gas (O_2). This reaction fits the characteristics of a decomposition reaction, making option A the correct answer.

QUESTION 155

Answer: A

Explanation: In a double displacement reaction, the cations and anions of two different compounds exchange places, resulting in the formation of two new compounds. This type of reaction is also known as a metathesis or exchange reaction.

QUESTION 156

Answer: A

Explanation: During an exothermic reaction, energy is released in the form of heat. However, in many cases, the heat energy is quickly absorbed by the surrounding environment, such as the reaction vessel, the air, or any nearby objects, before it can be transferred to our skin and be felt as a sudden burst of heat. The heat transfer between the reaction system and the surroundings occurs rapidly, making it difficult for us to perceive the immediate increase in temperature. Thus, the correct answer is option A.

QUESTION 157

Answer: A

Explanation: pH is a measure of the concentration of hydrogen ions ($H+$) in a solution. As the pH decreases from 5 to 3, it indicates an increase in the concentration of $H+$ ions, making the solution more acidic. Option B refers to hydroxide ions ($OH-$), which are not directly related to pH.

QUESTION 158

Answer: C

Explanation: Data tables are used to organize and present scientific data in a clear and concise format, making it easier for others to interpret and analyze the results.

QUESTION 159

Answer: D

Explanation: Radioisotopes are used in medical imaging and cancer treatment because they emit detectable radiation, which allows for diagnostic purposes and targeted destruction of cancer cells.

QUESTION 160

Answer: C

Explanation: Statistical analysis is a crucial tool in chemistry research for drawing valid conclusions from experimental data and quantifying uncertainties in the results. It helps researchers determine the reliability of their findings, assess the significance of observed differences, and identify any potential experimental errors or outliers. By using statistical methods, chemists can make meaningful interpretations of their data and present their results with appropriate measures of uncertainty.

QUESTION 161

Answer: A

Explanation: In a redox reaction, there is a transfer of electrons between the reactants. In the burning of wood (combustion) in the presence of oxygen, the wood is oxidized (loses electrons), and oxygen is reduced (gains electrons). This reaction involves a transfer of electrons, making option A the correct answer.

QUESTION 162

Answer: B

Explanation: The modern periodic table is arranged based on the increasing order of atomic numbers. Atomic number represents the number of protons in an atom's nucleus, and it is the fundamental property that determines an element's chemical properties and its position in the periodic table.

QUESTION 163

Answer: C

Explanation: The observed four-fold increase in the rate of reaction when doubling the concentration of substance X indicates that the reaction rate is directly proportional to the square of the concentration of X. This behavior is characteristic of a second-order reaction, where the rate is proportional to $[X]^2$.

QUESTION 164

Answer: D

Explanation: At saturation, the maximum amount of solute (sugar) that can dissolve in the solvent (water) at a specific temperature has already dissolved. Any additional sugar will not dissolve, leading to the observed excess solid sugar at the bottom of the container.

QUESTION 165

Answer: B

Explanation: Control variables are kept constant in an experiment to eliminate their potential influence on the results. This allows researchers to isolate and observe the effects of the independent variable.

QUESTION 166

Answer: D

Explanation: To foster creativity and innovation in chemistry, teachers should encourage students to think critically, explore new ideas, and design their experiments. Allowing students the freedom to make decisions in their experimental design can stimulate their curiosity and lead to unique insights and discoveries. By providing an environment that nurtures creativity and open-mindedness, chemistry teachers can inspire their students to become independent thinkers and problem-solvers in the field.

QUESTION 167

Answer: A

Explanation: Universal indicator is a mixture of several indicators that changes color based on the pH of a solution, which is a measure of the concentration of hydrogen ions (H+). In acidic solutions, the concentration of H+ ions is high, causing the indicator to turn red. In basic solutions, the concentration of H+ ions is low, leading to a blue color. A neutral solution, with an equal concentration of H+ and hydroxide ions (OH-), results in a green color. The universal indicator is an effective tool to differentiate between acidic, basic, and neutral solutions based on their pH levels.

QUESTION 168

Answer: B

Explanation: Formulating a hypothesis is the initial step in the scientific method. It involves making an educated guess or prediction about the outcome of an experiment, which can then be tested through experimentation.

QUESTION 169

Answer: B

Explanation: The relationship between solubility and temperature varies for different solutes and solvents. While some solutes exhibit increased solubility with higher temperatures, others may show the opposite trend. The student's suggestion is not valid because it oversimplifies the complex solubility behavior observed in different chemical systems.

QUESTION 170

Answer: C

Explanation: Addressing potential bias in scientific research is essential in chemistry and all scientific disciplines. Bias can arise at various stages of the research process, such as in experimental design, data collection, and interpretation of results. It can distort the outcomes and conclusions of an investigation. By recognizing and acknowledging potential sources of bias, researchers can take measures to mitigate its effects, improving the validity and reliability of their findings.

QUESTION 171

Answer: B

Explanation: Exergonic reactions are spontaneous reactions that release energy, usually in the form of heat. These reactions have a negative change in free energy ($\Delta G < 0$), indicating that the products have lower energy than the reactants. Exergonic reactions do not require an external energy source to proceed; instead, they release energy during the reaction. On the other hand, endergonic reactions are non-spontaneous reactions that require an input of energy to proceed. These reactions have a positive change in free energy ($\Delta G > 0$), indicating that the products have higher energy than the reactants. Endergonic reactions absorb energy from the surroundings to drive the reaction forward. Biological processes often involve both exergonic and endergonic reactions. Exergonic reactions provide the energy needed to drive endergonic reactions, coupling the reactions together to maintain the overall energy balance within living organisms.

QUESTION 172

Answer: C

Explanation: The history of chemistry provides insights into the progress of scientific knowledge and the evolution of instruments and techniques used by early chemists. Understanding their limitations helps present-day chemists appreciate the advancements made in the field and the reliability of current methodologies.

QUESTION 173

Answer: C

Explanation: Controls are used to establish a baseline for comparison in an experiment. They help identify the effects of the independent variable by showing what would happen without the experimental treatment.

QUESTION 174

Answer: C

Explanation: To calculate the concentration of the unknown acid, the teacher needs the concentration of the standardized base solution (the solution added during the titration). The volume and concentration of the base solution, along with the stoichiometry of the reaction, are essential components for determining the concentration of the unknown acid.

QUESTION 175

Answer: C

Explanation: A homogenous mixture is one that has a uniform composition and the same properties throughout the entire mixture. In such mixtures, the substances are thoroughly mixed at the molecular or atomic level, creating a uniform distribution of components. Option A (visibly distinct phases) and Option D (scattering light, making the mixture opaque) describe characteristics of heterogeneous mixtures, where the components are not uniformly distributed, leading to visible separations or scattering of light. Option B (variable composition throughout) also suggests a heterogeneous mixture, making option C the correct answer.

QUESTION 176

Answer: B

Explanation: Dalton's Atomic Theory, proposed by John Dalton in the early 19th century, suggested that atoms were indivisible and indestructible particles. According to this model, atoms were considered the basic building blocks of matter and did not contain any subatomic particles like electrons, protons, or neutrons.

QUESTION 177

Answer: D

Explanation: Element X, located in Group 17 of the periodic table, is in the halogen group. Halogens are known for their strong electron affinity, which means they have a high tendency to gain an electron to achieve a stable electron configuration with a complete outer electron shell (resembling the noble gas configuration). In this case, Element X has seven valence electrons in its electron configuration, and by gaining one more electron, it can achieve the stable configuration of the noble gas krypton (Kr). As a result, Element X is likely to form ions with a charge of -1, making it a halogen.

QUESTION 178

Answer: A

Explanation: Isotopes are variants of an element that have the same number of protons (which defines the element) but different numbers of neutrons, resulting in different mass numbers. This difference in the number of neutrons gives rise to isotopes, which may have slightly different physical properties but share similar chemical behavior due to the same number of protons and electrons.

QUESTION 179

Answer: C

Explanation: Flammability is a chemical property that describes how easily a substance will burn or ignite. Melting point (a), Color (b), and Density (d) are physical properties.

QUESTION 180

Answer: A

Explanation: According to Le Chatelier's principle, increasing the temperature of an equilibrium reaction will shift the equilibrium position in the direction of the endothermic reaction, in this case, the forward reaction.

QUESTION 181

Answer: A

Explanation: In an exothermic reaction, energy is released, causing the enthalpy of the products to be lower than that of the reactants.

QUESTION 182

Answer: A

Explanation: The electrical charge of an atom is determined by the difference between the number of protons (positive charge) and the number of electrons (negative charge). In this case, there are 12 protons (positive) and 11 electrons (negative). Thus, the atom has an overall positive charge of +1.

QUESTION 183

Answer: C

Explanation: According to Charles's Law, at constant pressure, the volume of an ideal gas is directly proportional to its temperature. When the temperature is doubled, the volume of the gas will also double. Therefore, the volume will be quadrupled (i.e., becomes four times larger) compared to its initial volume.

QUESTION 184

Answer: D

Explanation: Fluorine has the highest electronegativity among the given elements. Electronegativity is a measure of an atom's ability to attract and hold electrons in a chemical bond. Fluorine, being the most electronegative element, has a strong attraction for electrons.

QUESTION 185

Answer: A

Explanation: The Periodic Table is arranged in order of increasing atomic number (a), which is the number of protons in an atom's nucleus. Atomic mass (b) is not the primary factor for the arrangement, and options (c) and (d) are essentially the same since the number of protons and neutrons is nearly equal in stable atoms. 0006 Understand the kinetic molecular theory, the nature of phase changes, and the gas laws:

QUESTION 186

Answer: A

Explanation: A negative value of ΔG indicates that the reaction is spontaneous under standard conditions (constant temperature and pressure).

QUESTION 187

Answer: C

Explanation: Fluorine (F) has the highest electronegativity value among the elements listed. Electronegativity is a measure of an atom's ability to attract and hold electrons when it is chemically bonded to another atom. Fluorine, being a highly electronegative element, has a strong attraction for electrons and tends to pull them towards itself in a chemical bond.

QUESTION 188

Answer: A

Explanation: Boyle's Law states that at constant temperature, the pressure and volume of a given amount of gas are inversely proportional to each other. If the volume of the gas increases, its pressure will decrease, and vice versa, as long as the temperature remains constant.

QUESTION 189

Answer: A

Explanation: The atomic number of an element represents the number of protons in the nucleus of an atom of that element. Since the number of protons uniquely identifies an element, the atomic number is used to categorize elements in the periodic table.

QUESTION 190

Answer: A

Explanation: The First Law of Thermodynamics, also known as the Law of Energy Conservation, states that energy in a closed system remains constant; it can neither be created nor destroyed, only transformed from one form to another.

QUESTION 191

Answer: A

Explanation: Even if both reactions have the same negative ΔG, the reaction with the lower activation energy (Reaction A) will proceed faster and is more likely to occur spontaneously at room temperature compared to the one with the higher activation energy (Reaction B).

QUESTION 192

Answer: C

Explanation: A decrease in volume (negative work done by the system on the surroundings) and an increase in temperature suggest that the reaction releases heat to the surroundings, indicating an exothermic reaction. The negative work indicates that work is done by the system on the surroundings.

QUESTION 193

Answer: A

Explanation: Isotopes are variants of an element that have the same number of protons and electrons but different numbers of neutrons. Since they have the same number of protons and electrons, isotopes have identical chemical properties, but their atomic masses may differ due to the varying number of neutrons.

QUESTION 194

Answer: B

Explanation: The electron configuration of nitrogen (N) with atomic number 7 is 1s2 2s2 2p5. This indicates that there are two electrons in the 1s orbital, two electrons in the 2s orbital, and five electrons in the 2p orbital.

QUESTION 195

Answer: C

Explanation: Thermal energy is the internal energy associated with the random motion of atoms and molecules in a substance. It is related to temperature and determines the substance's overall heat content.

QUESTION 196

Answer: A

Explanation: Heat loss to the surroundings would result in less heat being measured by the calorimeter, leading to an underestimated value for the calculated ΔH of the reaction.

QUESTION 197

Answer: B

Explanation: The endothermic dissolution of Substance A suggests that breaking the intermolecular forces between the solute particles requires energy. Therefore, Substance A likely forms stronger intermolecular bonds with the solvent compared to Substance B, leading to an endothermic dissolution.

QUESTION 198

Answer: C

Explanation: Conformational isomers are isomers that have the same molecular formula and the same arrangement of atoms but differ in their spatial orientation due to rotation around a single bond.

QUESTION 199

Answer: D

Explanation: Hydrogen bonding is primarily responsible for the unique properties of water, such as high surface tension and specific heat capacity. Hydrogen bonds between water molecules create strong attractions that influence these properties.

QUESTION 200

Answer: A

Explanation: The poor conductivity of the resulting compound suggests that Element Y and Element Z likely formed a covalent bond. The compound's non-conducting behavior can be attributed to similar electronegativities of the elements, leading to a non-polar covalent bond. In non-polar covalent compounds, electrons are shared equally, resulting in no charged species to carry an electric current.

QUESTION 201

Answer: C

Explanation:

Benzene exhibits resonance, indicated by the presence of alternating single and double bonds between carbon atoms. In reality, the carbon-carbon bonds are not strictly single or double, but rather a hybrid of the two, resulting in a more stable and delocalized electron system.

QUESTION 202

Answer: B

Explanation: The reaction involves the decrease in the number of moles (2 moles of water vapor to 1 mole of hydrogen gas and 1 mole of oxygen gas). Increasing the pressure will shift the equilibrium position towards the side with fewer moles of gas, which means it will shift towards H_2 and O_2.

QUESTION 203

Answer: C

Explanation: A buffer solution is composed of a weak acid and its conjugate base, or a weak base and its conjugate acid. To have a buffer with a pH of 5, we would want the pKa of the weak acid to be close to the desired pH (around 5) to ensure optimal buffering capacity. In this case, the combination in option c has a weak acid with a pKa of 6, which is close to pH 5, and its conjugate base with a pKb of 8, ensuring a good buffer for pH 5.

QUESTION 204

Answer: B

Explanation: An organic acid contains the carboxylic acid functional group, which is characterized by the presence of the -COOH group.

QUESTION 205

Answer: C

Explanation: Metallic bonds are formed between metal atoms and involve the pooling (delocalization) of valence electrons, creating a "sea of electrons" that holds the metal atoms together.

QUESTION 206

Answer: C

Explanation:

In a covalent bond, atoms share electrons to achieve a stable electron configuration. This sharing creates a localized electron density between the bonded atoms, resulting in a strong bond that holds the atoms together.

QUESTION 207

Answer: A

Explanation:

While both ammonia and methane have similar molecular shapes, ammonia contains a highly electronegative nitrogen atom. The hydrogen atoms in ammonia have a partial positive charge, making them susceptible to hydrogen bonding with other electronegative atoms (such as nitrogen or oxygen). Methane, on the other hand, lacks strongly electronegative atoms, and its hydrogen atoms are not conducive to hydrogen bonding.

QUESTION 208

Answer: B

Explanation: Decreasing the volume of the container will increase the pressure of the system. According to Le Châtelier's principle, the system will shift in the direction that reduces the number of moles of gas to relieve the increased pressure. In this reaction, the forward direction (formation of NH_3) involves fewer moles of gas than the reverse direction (N_2 and H_2), so the equilibrium will shift towards NH_3.

QUESTION 209

Answer: B

Explanation: A strong base is a substance that readily and completely dissociates in water, releasing a large number of hydroxide ions (OH^-). As a result, the concentration of OH^- ions is high in the solution, leading to a high pH value. Strong bases are also good conductors of electricity in solution due to the high concentration of ions.

QUESTION 210

Answer: C

Explanation: The reducing agent is the species that causes another species to undergo reduction by providing electrons. When the oxidation number of a species decreases, it means it gains electrons and is thus being reduced. Therefore, it is the reducing agent in the redox reaction.

QUESTION 211

Answer: D

Explanation: In metallic solids, positive ions are arranged in a regular pattern, and electrons are delocalized, creating a "sea of electrons" that allows for electrical conductivity and malleability.

QUESTION 212

Answer: C

Explanation: NH3 (ammonia) is most likely to exhibit hydrogen bonding because it contains a hydrogen atom bonded to a highly electronegative nitrogen atom, and there is a lone pair of electrons on the nitrogen atom, making it a suitable hydrogen bond donor and acceptor.

QUESTION 213

Answer: C

Explanation:

Hydrogen bonds are formed between a hydrogen atom of one molecule and an electronegative atom (like oxygen or nitrogen) of another molecule. In water, hydrogen bonds between water molecules contribute to its high surface tension, cohesion, and other unique properties.

QUESTION 214

Answer: A

Explanation: A large equilibrium constant (Kc) value, such as 100 in this case, indicates that at equilibrium, the concentration of products is significantly higher than the concentration of reactants, meaning the reaction heavily favors the products.

QUESTION 215

Answer: B

Explanation: Increasing the pressure will shift the equilibrium position towards the side with fewer moles of gas to relieve the increased pressure. In this reaction, the forward direction (formation of CH3OH) involves fewer moles of gas than the reverse direction (CO and H2), so the equilibrium will shift towards CH3OH.

QUESTION 216

Answer: B

Explanation: This assesses the understanding of oxidation and reduction, which are fundamental concepts in electrochemistry. Oxidation involves the loss of electrons by a species, while reduction involves the gain of electrons. The mnemonic "OIL RIG" can be used to remember this: Oxidation Is Loss, Reduction Is Gain.

QUESTION 217

Answer: B

Explanation: According to IUPAC rules, the compound with the molecular formula C4H10 is named butane.

QUESTION 218

Answer: C

Explanation: The strength of an ionic bond is influenced by the atomic size of the atoms involved. Larger ions with a lower charge density form stronger ionic bonds.

QUESTION 219

Answer: A

Explanation:

In a metallic bond, metal atoms share their valence electrons, forming a "sea" of delocalized electrons within the metal lattice. These mobile electrons can move freely throughout the lattice, allowing metals to conduct electricity.

QUESTION 220

Answer: A

Explanation: If Qc is greater than Kc, it means that the concentrations of products are higher and the concentrations of reactants are lower than their respective equilibrium values. The reaction will shift to the right (forward direction) to reduce the reaction quotient and reach equilibrium.

QUESTION 221

Answer: A

Explanation: The pOH is calculated as the negative logarithm (base 10) of the hydroxide ion concentration in Molarity (M). pOH = -log(OH^-). In this case, pOH = -log(1.0×10^{-5}) ≈ 5.

QUESTION 222

Answer: D

Explanation: Reduction involves the gain of electrons. In this reaction, Ag^+(aq) gains two electrons to form Ag(s), which means Ag^+(aq) is undergoing reduction.

QUESTION 223

Answer: B

Explanation: Electrochemical cells are devices that utilize redox reactions to either generate electrical energy (in galvanic cells) or use electrical energy to drive non-spontaneous reactions (in electrolytic cells). Option B correctly describes the operation of electrochemical cells.

QUESTION 224

Answer: B

Explanation: An exothermic chemical reaction is one that releases heat to the surroundings. During such a reaction, energy is given off in the form of heat, resulting in an increase in the surrounding temperature.

QUESTION 225

Answer: D

Explanation: In a double replacement reaction, two clear, colorless solutions are combined, resulting in the formation of a solid precipitate. This type of reaction involves the exchange of positive ions between the reactants, leading to the formation of new compounds.

QUESTION 226

Answer: A

Explanation: In a photochemical reaction, light energy is absorbed by reactant molecules, leading to increased reactivity and a higher reaction rate. The absorbed energy allows reactants to surmount the activation energy barrier more easily, leading to an increased rate of product formation compared to the reaction in the absence of light.

QUESTION 227

Answer: A

Explanation: To determine the volume of 0.2 M NaOH solution required to completely react with 25 mL of 0.1 M HCl, we can apply stoichiometry principles based on the balanced chemical equation: HCl + NaOH → NaCl + H2O. First, we calculate the moles of HCl in the 25 mL of 0.1 M HCl solution using the formula Moles = (Molarity) x (Volume in liters). This yields 0.0025 moles of HCl. Since the equation indicates a one-to-one mole ratio between HCl and NaOH, we require the same number of moles of NaOH to fully react with HCl. We then determine the volume of 0.2 M NaOH solution needed to provide these moles, employing the formula Volume = Moles / Molarity. This results in 0.0125 liters, which is equivalent to 12.5 mL when converted from liters to milliliters. Consequently, 12.5 mL of 0.2 M NaOH solution is necessary to react completely with 25 mL of 0.1 M HCl, making option A, 12.5 mL, the correct answer.

QUESTION 228

Answer: D

Explanation: In an irreversible chemical reaction, the reactants are converted into products, and once this conversion occurs, the reaction cannot be reversed. The products formed cannot react to reform the original reactants, and the reaction proceeds in one direction only.

QUESTION 229

Answer: A

Explanation: Catalysts are not consumed during a reaction, and they should maintain their effectiveness throughout the entire process. If the catalyst becomes unstable or degrades over time, it can lead to a decrease in its catalytic activity, causing the reaction rate to slow down despite its initial effectiveness.

QUESTION 230

Answer: C

Explanation: Mass percent is calculated as (mass of solute / total mass of solution) x 100%. In this case, it would be (20 g / 120 g) x 100% = 16.67%.

QUESTION 231

Answer: B

Explanation: The mole is defined as the amount of substance that contains the same number of entities (atoms, molecules, ions, etc.) as there are atoms in exactly 12 grams of carbon-12. This number is Avogadro's number, approximately 6.022 x 10^23. Option b) correctly defines the mole as the mass of 6.022 x 10^23 atoms or molecules of a substance.

QUESTION 232

Answer: A

Explanation: Galvanizing is an electrochemical process used to protect iron and steel from corrosion. It involves coating the metal surface with a layer of zinc through a galvanic cell, creating a protective barrier against oxidation and rust.

QUESTION 233

Answer: C

Explanation: The rusting of iron (Fe) involves the loss of electrons by iron, which is oxidized, and the gain of electrons by oxygen (from oxygen molecules in the air), which is reduced. This transfer of electrons defines a redox (reduction-oxidation) reaction.

QUESTION 234

Answer: C

Explanation: In diffusion-controlled reactions, the rate is determined by how quickly reactant particles can come into contact with each other. Larger particles have a smaller surface area-to-volume ratio, which means they have less exposed surface area for collisions. Smaller particles, on the other hand, increase the rate of surface area-dependent reactions due to more exposed surface area for reactions to occur.

QUESTION 235

Answer: C

Explanation: At STP, 1 mole of any gas occupies 22.4 liters. First, calculate the moles of H2 from the given volume (5 liters) at STP. Then, using the mole ratio from the balanced equation, determine the moles of O2 required. Finally, convert moles of O2 to volume using the molar volume at STP.

QUESTION 236

Answer: C

Explanation: Option C is the most appropriate first step as it allows students to express their initial understanding and exposes potential misconceptions. By discussing their answers in pairs, students engage in peer-to-peer dialogue, which can help clarify misconceptions before moving on to more complex activities.

QUESTION 237

Answer: C

Explanation: Mr. Lee's objective is to assess students' participation and understanding of a laboratory experiment on chemical reactions and gather both qualitative and quantitative data. The most appropriate assessment method for this purpose is using a portfolio approach. Portfolios can include laboratory reports, reflections, and other artifacts that provide qualitative insights into students' understanding and skills. Additionally, quantitative data can be collected by evaluating the quality of their reports, accuracy in recording data, and application of theoretical concepts. Self-assessment surveys (option a) might provide insights, but they may not yield as much detailed information about the actual performance in the specific experiment. Standardized tests (option b) are not tailored to this specific experiment, and a group project (option d) might not capture individual understanding and performance accurately.

QUESTION 238

Answer: B

Explanation: In a precipitation reaction, two solutions of different substances are mixed, resulting in the formation of a solid precipitate. This occurs when the cations and anions of the reactants combine to form an insoluble compound, which then falls out of solution as a solid.

QUESTION 239

Answer: C

Explanation: Option C involves hands-on experimentation and allows students to make systematic observations and measurements of reactants and products in a chemical reaction. This practical approach helps students connect theoretical stoichiometric calculations with real-world measurements, leading to a deeper understanding of the topic.

QUESTION 240

Answer: B

Explanation: Providing audio recordings of the texts can be a useful strategy to accommodate English-language learners and students with special needs who may benefit from auditory learning. It supports their comprehension by presenting the information in a format that suits their learning preferences. This approach respects student diversity and ensures that all students have access to the content-related texts in a way that aligns with their abilities and learning styles.

QUESTION 241

Answer: C

Explanation: Option C facilitates building upon students' prior knowledge by discussing the significance of chemical reactions in everyday life. This approach helps students relate the new concept of chemical reactions to their experiences, making it more meaningful and relatable.

QUESTION 242

Answer: C

Explanation: Fostering collaboration among students can be achieved through a group experiment where they work together to achieve a common goal. This activity not only encourages interaction among students but also promotes teamwork, communication, and problem-solving skills. Collaborative activities like this align with the importance of planning strategies to encourage student self-motivation and engagement in their own learning, enhancing their overall scientific understanding and learning experience.

QUESTION 243

Answer: D

Explanation: Option D involves hands-on experimentation and allows students to make systematic observations and measurements, such as temperature changes during chemical reactions. This practical approach helps students understand the concept of chemical reactions in a more tangible and concrete way.

QUESTION 244

Answer: C

Explanation: Option C aligns with the strategy of assisting students in generating, refining, focusing, and testing scientific s and hypotheses. By conducting brainstorming sessions, students can freely explore various ideas and interests related to the project topic. The teacher's guidance during these sessions helps students refine their s, ensuring they are meaningful, relevant, and aligned with the scientific investigation.

QUESTION 245

Answer: B

Explanation: Option B involves an inquiry-based approach where students are given a task to identify the type of chemical bond in each compound. This encourages higher-level thinking skills and logical reasoning as students have to analyze the properties of each compound to determine the bond type.

QUESTION 246

Answer: B

Explanation: In a precipitation reaction, the reactants form an insoluble product (precipitate). The balanced chemical equation is $AgNO_3 + NaCl ->AgCl + NaNO_3$. To find the mass of AgCl formed, we first need to find the limiting reactant. Moles of $AgNO_3$ = 0.2 M * 0.2 L = 0.04 mol. Moles of NaCl = 0.15 M * 0.15 L = 0.0225 mol. The stoichiometric ratio between $AgNO_3$ and AgCl is 1:1, so the limiting reactant is NaCl. Moles of AgCl = 0.0225 mol. Mass of AgCl = moles * molar mass = 0.0225 mol * 143.32 g/mol = 3.225 g. However, we only need to consider the mass of AgCl formed in the reaction, so we multiply by the ratio of volume used: 3.225 g * (200 mL / 350 mL) ≈ 0.24 g.

QUESTION 247

Answer: B

Explanation: Ammonia (NH_3) has a trigonal pyramidal shape due to its three bonded hydrogen atoms and a lone pair of electrons on the nitrogen atom.

QUESTION 248

Answer: B

Explanation: Iron (Fe) exhibits metallic bonding since it is a metal and has delocalized electrons that are free to move throughout the metal lattice, allowing it to conduct electricity.

QUESTION 249

Answer: C

Explanation: When the volume of a system decreases at constant pressure, work is done on the system by the surroundings. In an exothermic reaction, the system releases heat (q) to the surroundings. As a result, the enthalpy change (ΔH) for the reaction is positive, indicating that the reaction is endothermic. However, since the heat is being released to the surroundings, q is negative.

QUESTION 250

Answer: B

Explanation: Option B involves hands-on experimentation, which encourages higher-level thinking skills and logical reasoning. By testing the pH of different substances, students can observe patterns and draw conclusions about whether they are acids or bases, fostering deeper understanding and critical thinking.

QUESTION 251

Answer: C

Explanation: To evaluate the reliability and validity of a new assessment instrument, the teacher should compare the results obtained from the new assessment with results from a well-established and reliable assessment that measures the same learning objectives. This process, known as criterion-related validity, allows the teacher to assess the extent to which the new assessment measures what it intends to measure in comparison to a trusted benchmark. The other options (A, B, and D) do not directly address the evaluation of reliability and validity.

QUESTION 252

Answer: A

Explanation: Molarity (M) is defined as moles of solute per liter of solution. Solution X has 1.5 moles of solute X in 0.5 L of water, so its molarity is 1.5 mol / 0.5 L = 3 M. Solution Y has 2 moles of solute Y in 1 L of water, so its molarity is 2 mol / 1 L = 2 M. Therefore, solution X has a higher concentration

QUESTION 253

Answer: C

Explanation: Option C encourages higher-level thinking skills and scientific problem-solving by presenting students with a real-world problem. This approach allows students to apply their knowledge and understanding of chemical reactions to propose a solution. It moves them beyond memorization and requires them to think critically and logically.

QUESTION 254

Answer: A

Explanation: Corrosive chemicals can cause severe skin and eye damage. Wearing gloves and a lab coat provides protection from potential spills and splashes, ensuring the safety of the experimenter.

QUESTION 255

Answer: B

Explanation: An element is a pure substance composed of only one type of atom. It cannot be broken down into simpler substances by chemical reactions.

QUESTION 256

Answer: A

Explanation: Solubility refers to the ability of a solute (such as sugar) to dissolve in a solvent (such as water or orange juice) to form a homogeneous solution. The observation can be explained by the fact that sugar molecules have a higher solubility in water than in orange juice. This means that more sugar molecules can effectively interact with the water molecules, leading to faster dissolution in water compared to orange juice.

QUESTION 257

Answer: C

Explanation: The most likely cause for the observed differences in reaction rates is the temperature of the reaction environment. Increasing the temperature of a chemical reaction typically increases the rate of the reaction. Higher temperatures lead to increased kinetic energy of the reactant particles, which results in more frequent and energetic collisions between the reacting molecules, thus accelerating the reaction rate.

QUESTION 258

Answer: C

Explanation: The atomic number of an element is equal to the number of protons in the nucleus. It uniquely identifies the element and determines its chemical properties, as the number of protons defines the element's chemical behavior.

QUESTION 259

Answer: A

Explanation: According to the law of conservation of mass, mass is neither created nor destroyed during a chemical reaction. Therefore, the total mass of the reactants (A and B) will be equal to the total mass of the products (X and Y).

QUESTION 260

Answer: C

Explanation: Electron configuration notation represents the distribution of electrons in an atom's energy levels and sublevels. It uses numbers and letters to indicate the number of electrons in each energy level and the sublevel they occupy (e.g., $1s^2\ 2s^2\ 2p^6$ for neon). This notation provides essential information about the electron arrangement in an atom.

QUESTION 261

Answer: B

Explanation: While pH values are indicators of the relative acidity or basicity of a substance, they do not solely determine the strength of an acid or a base. The strength of an acid or base is related to its ability to donate or accept protons (H+ ions) in a chemical reaction. For example, hydrochloric acid (pH 1) is stronger than acetic acid (vinegar, pH 3) because it dissociates more readily in water, releasing more H+ ions. Similarly, ammonia solution (pH 11) is a weaker base than hydroxide solution (pH 14) because it accepts fewer H+ ions. Thus, factors other than pH must be considered when determining the strength of an acid or base.